CONTEMPORARY MATHEMATICS

Titles in this Series

VOLUME 1 **Markov random fields and their applications**
Ross Kindermann and J. Laurie Snell

VOLUME 2 **Proceedings of the conference on integration, topology, and geometry in linear spaces**
William H. Graves, Editor

VOLUME 3 **The closed graph and P-closed graph properties in general topology**
T. R. Hamlett and L. L. Herrington

VOLUME 4 **Problems of elastic stability and vibrations**
Vadim Komkov, Editor

VOLUME 5 **Rational constructions of modules for simple Lie algebras**
George B. Seligman

VOLUME 6 **Umbral calculus and Hopf algebras**
Robert Morris, Editor

VOLUME 7 **Complex contour integral representation of cardinal spline functions**
Walter Schempp

VOLUME 8 **Ordered fields and real algebraic geometry**
D. W. Dubois and T. Recio, Editors

VOLUME 9 **Papers in algebra, analysis and statistics**
R. Lidl, Editor

VOLUME 10 **Operator algebras and K-theory**
Ronald G. Douglas and Claude Schochet, Editors

VOLUME 11 **Plane ellipticity and related problems**
Robert P. Gilbert, Editor

VOLUME 12 **Symposium on algebraic topology in honor of José Adem**
Samuel Gitler, Editor

Titles in this series

VOLUME 13 **Algebraists' homage: Papers in ring theory and related topics** Edited by S. A. Amitsur, D. J. Saltman and G. B. Seligman

VOLUME 14 **Lectures on Nielsen fixed point theory** Boju Jiang

VOLUME 15 **Advanced analytic number theory. Part I: Ramification theoretic methods** Carlos J. Moreno

VOLUME 16 **Complex representations of GL(2, K) for finite fields K** Ilya Piatetski-Shapiro

VOLUME 17 **Nonlinear partial differential equations** Joel A. Smoller, Editor

VOLUME 18 **Fixed points and nonexpansive mappings** Robert C. Sine, Editor

VOLUME 19 **Proceedings of the Northwestern homotopy theory conference** Haynes R. Miller and Stewart B. Priddy, Editors

VOLUME 20 **Low dimensional topology** Samuel J. Lomonaco, Jr., Editor

VOLUME 21 **Topological methods in nonlinear functional analysis** Edited by S. P. Singh, S. Thomeier, and B. Watson

VOLUME 22 **Factorizations of $b^n \pm 1, b = 2, 3, 5, 6, 7, 10, 11, 12$ up to high powers** John Brillhart, D. H. Lehmer, J. L. Selfridge, Bryant Tuckerman, and S. S. Wagstaff, Jr.

VOLUME 23 **Chapter 9 of Ramanujan's second notebook—Infinite series identities, transformations, and evaluations** Bruce C. Berndt and Padmini T. Joshi

VOLUME 24 **Central extensions, Galois groups, and ideal class groups of number fields** A. Fröhlich

CONTEMPORARY MATHEMATICS

Volume 24

Central Extensions, Galois Groups, and Ideal Class Groups of Number Fields

A. Fröhlich

AMERICAN MATHEMATICAL SOCIETY
Providence · Rhode Island

1980 *Mathematics Subject Classification.* Primary 12Axx, 12Bxx.

Library of Congress Cataloging in Publication Data

Fröhlich, A. (Albrecht), 1916–

Central extensions, Galois groups, and ideal class groups of number fields.
(Contemporary mathematics, ISSN 0271-4132; v. 24)
Bibliography: p.
1. Class field theory. 2. Field extensions (Mathematics) 3. Galois theory. 4. Class groups (Mathematics) I. Title. II. Series: Contemporary mathematics (American Mathematical Society); v. 24)

QA247.F758 1983 512′.3 83-19685
ISBN 0-8218-5022-9

This volume was printed directly from author-prepared copy.
The paper used in this book is acid-free and falls within the guidelines
established to ensure permanence and durability.

10 9 8 7 6 5 4 3 2 95 94 93 92 91 90

Table of Contents

PREFACE

These notes deal with a set of interrelated problems and results in algebraic number theory, in which there has been renewed activity in recent years. The underlying tool is the theory of the central extensions and, in vaguest and most general terms, the underlying aim is to use class field theoretic methods to reach beyond Abelian extensions. The literature in this area is now quite extensive, with often differing but overlapping approaches. One purpose of this write-up is to give an introductory survey, assuming the basic theorems of class field theory, as mostly recalled in §1. No originality is claimed as regards the general theory, although our approach seems to have some novel features, in particular in the central role that the Tate cohomology groups $\hat{H}^{-1}$ play.

Our principal aim is however to use the general theory, as developed here, together with the special features of class field theory over $\mathbb{Q}$, to derive some rather strong theorems of a very concrete nature, with $\mathbb{Q}$ as basefield. This is where the main emphasis of these notes lies. The specialization of the theory of central extensions to the base field $\mathbb{Q}$ will be shown to derive from an underlying principle of wide applicability. We shall describe certain non Abelian Galois groups over the rational field, and their inertia subgroups and use this description to gain information on ideal class groups of absolutely Abelian fields, and this in entirely rational terms. Here indeed we shall obtain precise and explicit arithmetic results, which go far beyond anything available in the general theory. Much of this really goes back to the author's thesis, but has so far not been genuinely accessible in terms of modern class field theory, and indeed our proofs are entirely new. As we shall show, some of the results on class groups have also an interpretation in terms of the group of norms modulo total norm residues, the so-called number knot.

The theory of the genus field, which is needed as background, but of course is of independent interest, is presented in §2. Then we develop the theory of central extension in §3. The special features over $\mathbb{Q}$ are pointed out throughout, also in §1. Next, §4 deals with Galois groups, and in §5 come the applications to class groups. Finally §6 contains some remarks on the history and literature, but no completeness is attempted.

Acknowledgement. This survey was written while I was visiting George A. Miller Professor at the University of Illinois at Urbana-Champaign, during the special Algebra year 1981/82, organized by I. Reiner. It is based on some lectures, given during this period. My thanks are due to the University of Illinois for its support and to my colleagues in the Mathematics Department there, for their hospitality and for listening to me.

§1. Background from Class Field Theory

(a) Profinite groups. A profinite group G is the inverse limit of finite groups with the inverse limit topology. Thus G is a compact, totally disconnected group; it has a system of open neighborhoods of the identity consisting of normal subgroups of finite index.

The standard example is that of a Galois group of a Galois extension E/F of fields. Here

$$\mathrm{Gal}(E/F) = \varprojlim \mathrm{Gal}(E_i/F),$$

with E_i/F running over the Galois subextensions of E/F of finite degree.

The inclusion of the category of profinite groups in that of topological groups and continuous homomorphisms has a left adjoint $G \longmapsto \tilde{G}$. Thus given a topological group G there is a profinite group $\tilde{G}$ and a continuous homomorphism $G \to \tilde{G}$, such that every continuous homomorphism $G \to H$, with H profinite, factorises uniquely as in the commutative diagram

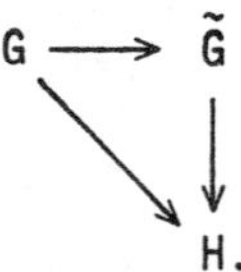

Explicitly

$$\tilde{G} = \varprojlim G/N, \tag{1.1}$$

with N running through the closed (hence open) normal subgroups of finite index.

If G is Abelian, then so is $\tilde{G}$. Write $\check{G}$ for the Pontryagin dual, and

$G^\dagger$ for the torsion subgroup of $\check{G}$. Character "restriction" from $\tilde{G}$ to G yields an isomorphism

$$\check{\tilde{G}} \simeq G^\dagger, \tag{1.2}$$

as one easily verifies.

<u>Examples</u>. (As usual the symbols $\mathbb{Z}$, $\mathbb{Q}$, $\mathbb{Q}_p$, $\mathbb{R}$, $\mathbb{C}$ stand for the ring of integers, and the fields of rational, p-adic rational, real, and complex numbers, respectively. $S^\times$ is the multiplicative group of units of a ring S.)

1) If G is a discrete group, give it a new topology by taking the normal subgroups of finite index as a system of open neighborhoods of the identity. Then complete and you get $\tilde{G}$. The case $\tilde{\mathbb{Z}}$ is well known.

2) We take $\mathbb{R}^\times$, $\mathbb{C}^\times$ with their natural topologies. Then $\tilde{\mathbb{R}}^\times = \pm 1$, $\tilde{\mathbb{C}}^\times = 1$.

3) Let F be a local field, which here will always mean an extension of finite degree of $\mathbb{Q}_p$, for some p, or of $\mathbb{R}$. Suppose F is non Archimedean, i.e., not $\mathbb{R}$ or $\mathbb{C}$. Write $U(F)$ for the group of units of the ring of integers of F. Then $F^\times \simeq U(F) \times \mathbb{Z}$, and $\tilde{F}^\times \cong U(F) \times \tilde{\mathbb{Z}}$.

4) Let K be a number field, i.e., an extension of $\mathbb{Q}$ of finite degree. Let $J(K)$ be its idele group, and embed $K^\times$ diagonally in $J(K)$. Then $C(K) = J(K)/K^\times$ is the idele class group. ($K^\times$ is discrete in $J(K)$). Let now $D(K)$ be the connected component of the identity in $C(K)$. Then

$$\tilde{C}(K) = C(K)/D(K). \tag{1.3}$$

Let $U(K)$ be the group of unit ideles, i.e., the product $\Pi_{\mathfrak{p}} U(K_{\mathfrak{p}})$, where $\mathfrak{p}$ runs through all prime divisors of K, and $K_{\mathfrak{p}}$ is the completion of K at $\mathfrak{p}$. If $\mathfrak{p}$ is finite, we have already defined $U(K_{\mathfrak{p}})$. If $K_{\mathfrak{p}} = \mathbb{R}$ or $= \mathbb{C}$, then $U(K_{\mathfrak{p}}) = K_{\mathfrak{p}}^*$. For finite $\mathfrak{p}$, $\widetilde{U(K_{\mathfrak{p}})} = U(K_{\mathfrak{p}})$, and in the other case we go back to example 2. Then $\widetilde{U(K)} = \Pi_{\mathfrak{p}} \widetilde{U(K_{\mathfrak{p}})}$.

If H is a subgroup of $J(K)$, we denote by H_+ the subgroup of H of totally positive elements x in H, i.e., of elements x with $x_{\mathfrak{p}} > 0$ whenever $\mathfrak{p}$ is real. Let $U_{fin}(K)$ be the subgroup of $U(K)$ of unit ideles with

$x_{\mathfrak{p}} = 1$ whenever $\mathfrak{p}$ is infinite. Then $U(K)_+$ is the direct product of $U_{fin}(K)$, of copies of $\mathbb{C}^*$ -- one for each complex prime $\mathfrak{p}$ of K -- and of copies of the group $\mathbb{R}_+$ of positive reals -- one for each real prime $\mathfrak{p}$ of K. Now we get

$$\widetilde{U(K)}_+ = U_{fin}(K). \tag{1.4}$$

(b) The reciprocity map. Here, to begin with, K is a number field, or a local field. L is a Galois extension of K with Galois group Γ. If L is a number field write $A(L) = C(L)$, if it is a local field, write $A(L) = L^*$. Class field theory gives us then canonical isomorphisms

$$\hat{H}^n(\Gamma, \mathbb{Z}) \cong \hat{H}^{n+2}(\Gamma, A(L)), \quad \text{for all} \quad n \in \mathbb{Z}, \tag{1.5}$$

where the $\hat{H}^n$ are the Tate cohomology groups. Observing that $\hat{H}^{-2}(\Gamma, \mathbb{Z}) = \Gamma^{ab}$ and taking the inverse of the isomorphism (1.5) for $n = -2$, we get an exact sequence

$$1 \to N_{L/K}(A(L)) \to A(K) \to \Gamma^{ab} \to 1, \tag{1.6}$$

where $N_{L/K}$ always denotes the norm operator. Hence

$$N_{L/K}(A(L)) = N_{L'/K}(A(L')), \quad L' \text{ the maximal Abelian extension of } K \text{ in } L. \tag{1.7}$$

The sequences (1.6) behave well with respect to variation of the top field L. Going to the limit over Abelian extensions L of K, we get accordingly a reciprocity map

$$A(K) \to \mathrm{Gal}(K^{ab}/K), x \longmapsto (K,x), \tag{1.8}$$

K^{ab} the maximal Abelian extension of K. If $K \subset M \subset K^{ab}$ we shall write the resulting composite

$$A(K) \to \mathrm{Gal}(K^{ab}/K) \to \mathrm{Gal}(M/K)$$

as

$$x \longmapsto (M/K,x).$$

We shall also need one of the maps associated with a change of base field.

Let K' be an extension of K of finite degree. Then $K'^{ab} \supset K^{ab}$, and we have a commutative diagram

$$\begin{array}{ccc} A(K') & \longrightarrow & \mathrm{Gal}(K'^{ab}/K') \\ {\scriptstyle N_{K'/K}}\downarrow & & \downarrow \\ A(K) & \longrightarrow & \mathrm{Gal}(K^{ab}/K), \end{array} \tag{1.9}$$

the right-hand column being the canonical map of Galois groups.

From now on let K be a number field. The global and the local maps are then related by a commutative diagram

$$\begin{array}{ccc} K_{\mathfrak{p}}^{\times} & \longrightarrow & \mathrm{Gal}(K_{\mathfrak{p}}^{ab}/K_{\mathfrak{p}}) \\ \downarrow & & \downarrow \\ C(K) & \longrightarrow & \mathrm{Gal}(K^{ab}/K) \end{array} \tag{1.10}$$

the left-hand column being the composite $K_{\mathfrak{p}}^{\times} \to J(K) \to C(K)$.

At the finite level, for an Abelian extension L of K of finite degree, we get analogously a commutative diagram

$$\begin{array}{ccc} K_{\mathfrak{p}}^{\times} & \longrightarrow & \mathrm{Gal}(L_{\mathfrak{P}}/K_{\mathfrak{p}}) \\ \downarrow & & \downarrow \\ C(K) & \longrightarrow & \mathrm{Gal}(L/K), \end{array} \tag{1.10a}$$

where $\mathfrak{P}$ is some prime divisor of L above $\mathfrak{p}$. Via local class field theory we then deduce that for finite $\mathfrak{p}$

$$U(K_{\mathfrak{p}}) \subset \mathrm{Ker}[K_{\mathfrak{p}}^{\times} \to \mathrm{Gal}(L_{\mathfrak{P}}/K_{\mathfrak{p}})] \Leftrightarrow \mathfrak{p} \text{ is non-ramified in } L. \tag{1.11}$$

For infinite $\mathfrak{p}$ we shall take this as the definition of "non-ramified".

The <u>norm subgroups</u> $N_{L/K}(C(L))$, L/K of finite degree as before, are closed of finite index in $C(K)$. Conversely one also knows that every such subgroup H of $C(K)$ is the norm subgroup for a (unique!) Abelian extension L/K; we shall then call L the <u>class field</u> of $C(K)/H$ and $C(K)/H$ the <u>class</u>

<u>group</u> associated with L. Note: Whenever we speak of a class group $\mathfrak{C}$ we mean by this a group explicitly given as a quotient of $C(K)$. In other words, a class group is really a diagram $C(K) \to \mathfrak{C}$. Very often we shall describe such a class group as a quotient $J(K)/G$ with $G \supset K^\times$.

The reciprocity maps

$$\begin{cases} C(K) \to \mathrm{Gal}(K^{ab}/K) & \text{(global case)}, \\ F^\times \to \mathrm{Gal}(F^{ab}/F) & \text{(local case)}, \end{cases} \tag{1.12}$$

are continuous, but are not isomorphisms. This is what lies behind the introduction of global or local Weil groups, which for the particular extensions K^{ab}/K and F^{ab}/F give a precise version of (1.12). For our purpose a more simple minded approach is more useful. Observe that (1.12) gives rise to continuous homomorphisms

$$\begin{cases} \widetilde{C(K)} \to \mathrm{Gal}(K^{ab}/K), \\ \widetilde{F^\times} \to \mathrm{Gal}(F^{ab}/F). \end{cases} \tag{1.12a}$$

One then shows that these are in fact isomorphisms. It now follows from this and (1.2) that the norm subgroups and their relationships can be described in terms of $C(K)^\dagger$, i.e., of idele class characters ϕ of finite order. If L is a Galois extension of K of finite degree, the character group $\Phi(L/K)$ of $C(K)/N_{L/K}(C(L))$ can, by lifting, be viewed as a subgroup of $C(K)^\dagger$. In this way we obtain a bijection between the set of Abelian extensions L/K of finite degree inside a given algebraic closure, and the set of finite subgroups of $C(K)^\dagger$ (i.e., of $\check{C}(K)$). There is a slightly different way for looking at this. The norm map $N_{L/K}$ is continuous. We thus get a conorm $\check{N}_{L/K}$, which maps: $C(K)^\dagger \to C(L)^\dagger$. Then we have

$$\Phi(L/K) = \mathrm{Ker}\ \check{N}_{L/K}. \tag{1.13}$$

Analogous definitions apply in the local case. We shall also view a character ϕ of $C(K)$ as a character of $J(K)$, trivial on $K^\times$. Its local component $\phi_{\mathfrak{p}}$

at a prime $\mathfrak{p}$ of K is its restriction to $K_{\mathfrak{p}}^{\times}$. For such a character ϕ we then have the <u>product relation</u> $\Pi_{\mathfrak{p}}\phi_{\mathfrak{p}}(a) = 1$, for all $a \in K^{\times}$.

<u>Examples</u>. 1) Let $I(K)$ be the group of non-zero fractional ideals in K (i.e., of the ring of integers of K), $P(K)$ the subgroup of principal ideals. Then

$$\mathfrak{Cl}(K) = I(K)/P(K) \tag{1.14}$$

is the absolute ideal class group. We shall show that $\mathfrak{Cl}(K)$ can be viewed in a natural way as a class group in the sense defined earlier, by establishing a natural isomorphism

$$\mathfrak{Cl}(K) \cong J(K)/U(K)K^{\times}. \tag{1.15}$$

The class field associated with $\mathfrak{Cl}(K)$ is the <u>absolute class field of</u> K. By (1.11) it is the maximal Abelian extension of K, non-ramified over K.

To obtain (1.15) we define the <u>contents</u> (x) of an idele x as the fractional ideal with components $(x)_{\mathfrak{p}} = (x_{\mathfrak{p}})$ at all finite primes. The map $x \longrightarrow (x)$ is a surjective homomorphism $J(K) \to I(K)$, inducing a surjection $K^{\times} \to P(K)$. This then yields the required isomorphism. Equivalently we can describe the result by an exact sequence

$$1 \to Y(K) \to U(K) \to C(K) \to \mathfrak{Cl}(K) \to 1, \tag{1.15a}$$

where $Y(K) = U(K) \cap K^{\times}$ is the group of global units.

2) Let $P(K)_{+}$ be the group of totally positive principal ideals, i.e., the image of $K_{+}^{\times}$ in $I(K)$. The <u>narrow absolute ideal class group</u> is defined by

$$\mathfrak{Cl}_{+}(K) = I(K)/P(K)_{+}. \tag{1.16}$$

In the same way as we derived (1.15) from (1.14), we now get an isomorphism

$$\mathfrak{Cl}_{+}(K) \cong J(K)/U(K)K_{+}^{\times}. \tag{1.17}$$

This, however, does not exhibit $\mathfrak{Cl}_{+}(K)$ as a class group in the sense we have defined it, i.e., as a quotient of $C(K)$. But we shall derive an isomorphism

$$\mathfrak{Cl}_+(K) \cong J(K)/U(K)_+K^\times, \tag{1.18}$$

or equivalently an exact sequence

$$1 \to Y(K)_+ \to U(K)_+ \to C(K) \to \mathfrak{Cl}_+(K) \to 1. \tag{1.18a}$$

The class field of the class group (1.18) is the narrow absolute class field.

To obtain (1.18) we use (1.17) and define a surjective homomorphism

$$f\colon J(K) \to J(K)/U(K)_+K^\times,$$

with $U(K)K_+^\times$ as kernel. If $x \in J(K)$ then we put $f(x)$ = class mod $U(K)_+K^\times$ of $x \cdot x_\infty^{-1}$. Here $(x_\infty)_\mathfrak{p} = x_\mathfrak{p}$ if $\mathfrak{p}$ is infinite, $(x_\infty)_\mathfrak{p} = 1$ if $\mathfrak{p}$ is finite. One knows that all possible signatures (combinations of signs at the real primes) are taken by elements of K. Choose $a \in K^\times$, with the same signature as a given idele x. Then $f(xa^{-1})$ = class of x mod $U(K)_+K^\times$.

If $x \in K_+^\times$, then $x_\infty \in U(K)_+$, whence $x\,x_\infty^{-1} \in U(K)_+K^\times$, i.e., $f(x) = 1$. Again if $x \in U(K)$ then $x\,x_\infty^{-1} \in U(K)_+$, $f(x) = 1$. Conversely if $x \in \operatorname{Ker} f$, i.e., $x\,x_\infty^{-1} = yz$, $y \in K^\times$, $z \in U(K)_+$, then $y_\infty z_\infty = 1$, i.e., $y \in K_+^\times$. Also $x_\infty z \in U(K)$, hence $x \in U(K)K_+^\times$.

The map from $\mathfrak{Cl}_+(K)$ onto $\mathfrak{Cl}(K)$ fits into an exact sequence, involving the signature group

$$sign(K) = \operatorname{Hom}(\mathbb{F}_2\mathfrak{R}, \pm 1)$$

($\mathbb{F}_p$ is always the field of p elements), where $\mathfrak{R} = \mathfrak{R}(K)$ is the set of real primes of K. We have a map $sign\colon J(K) \to sign(K)$, where $sign(x)(\mathfrak{p}) = sign_\mathfrak{p}(x) = x_\mathfrak{p}/|x_\mathfrak{p}|$, for all $\mathfrak{p} \in \mathfrak{R}$. We have already noted that $sign\colon K^\times \to sign(K)$ is surjective. We now restrict the map $sign$ to $Y(K)$, and we define a homomorphism $sign(K) \to \mathfrak{Cl}_+(K)$ as follows. If $s \in sign(K)$ let $a \in K^\times$ have $sign(a) = s$. Then map s onto the class of the principal ideal (a). With these definitions and the inclusion map $Y(K)_+ \to Y(K)$ we get an exact sequence

$$1 \to Y(K)_+ \to Y(K) \to sign(K) \to \mathfrak{Cl}_+(K) \to \mathfrak{Cl}(K) \to 1. \tag{1.19}$$

For the proof we write down a second sequence

$$1 \to Y(K)_+ \to Y(K) \to U(K)/U(K)_+ \to J(K)/U(K)_+K^\times \to J(K)/U(K)K^\times \to 1, \tag{1.19a}$$

whose maps are the obvious ones as subquotients of $J(K)$. It is clearly exact. Moreover the map *sign* defines an isomorphism $U(K)/U(K)_+ \simeq sign(K)$. This, together with the inverse isomorphisms of (1.15), (1.18), yields an isomorphism of (1.19a) onto (1.19).

We now look at the dual of the map $U(K)_+ \to C(K)$ in (1.18a). By (1.4) we obtain a homomorphism

$$C(K)^\dagger \to U_{fin}(K)^\dagger = \check{U}_{fin}(K). \tag{1.20}$$

Its kernel is $\check{\mathfrak{Cl}}_+(K)$, and its image the subgroup of $\check{U}_{fin}(K)$ of characters trivial on the image of $Y(K)_+$.

<u>The rational case</u>. We clearly have $\mathfrak{Cl}_+(\mathbb{Q}) = 1$, $Y(\mathbb{Q})_+ = 1$ and so (1.20) is an isomorphism. Indeed we have a direct product

$$J(\mathbb{Q}) = U_{fin}(\mathbb{Q})\times\mathbb{Q}^\times\times \mathbb{R}_+^\times. \tag{1.21}$$

The isomorphism (1.20) is then given by the projection $J(\mathbb{Q}) \to U_{fin}(\mathbb{Q})$. Given a character ϕ of $U_{fin}(\mathbb{Q})$ we can extend it simply by letting it map $\mathbb{Q}^\times \times \mathbb{R}_+^\times$ onto 1. The infinite component of the resulting idele class character is given by $\phi_\infty(-1) = \phi((-1)_{fin})^{-1}$, $(-1)_{fin}$ the image of -1 in $U_{fin}(\mathbb{Q})$.

In the triangle

$$\begin{array}{ccc} U_{fin}(\mathbb{Q}) & \xrightarrow{\ a\ } & \widetilde{C(\mathbb{Q})} \\ {\scriptstyle c}\nwarrow & & \swarrow{\scriptstyle b} \\ & \mathrm{Gal}(\mathbb{Q}^{ab}/\mathbb{Q}) & \end{array} \tag{1.22}$$

we already have the isomorphism a from the structure of $J(\mathbb{Q})$, i.e., from (1.21), and the reciprocity isomorphism b. We shall now give a direct definition of an isomorphism c which makes the diagram anticommute.

Let μ be the group of all roots of unity (over $\mathbb{Q}$). Then

$$\mathrm{Aut}(\mu) \cong U_{fin}(\mathbb{Q}). \tag{1.23}$$

Let μ_{p_∞} be the group of all p-power roots of unity. Then clearly $\mathrm{Aut}(\mu_{p_\infty}) \cong U(\mathbb{Q}_p)(=\mathbb{Z}_p^\times)$, and we get (1.23) by taking the product. On the other hand, $\mathbb{Q}^{ab} = \mathbb{Q}(\mu)$ and so we get an injective homomorphism

$$\mathrm{Gal}(\mathbb{Q}^{ab}/\mathbb{Q}) \to \mathrm{Aut}(\mu) \tag{1.24}$$

by restricting the action to μ. This is actually an isomorphism, and c is the composite of (1.23), (1.24). One then indeed finds that

$$c \circ b \circ a(u) = u^{-1}.$$

§2. The Genus Field and the Genus Group

The genus group $\mathfrak{G}(L/K)$, to be defined here is a quotient of the absolute class group $\mathfrak{Cl}(L)$ associated with an extension L/K of number fields, and described entirely in terms of class field theory over the base field K. Analogously we shall also consider the "narrow" genus group $\mathfrak{G}_+(L/K)$. Although the theory is quite general, we shall restrict ourselves to Galois extensions L/K. When $K = \mathbb{Q}$ we shall be able to extract a good deal of information. In particular we shall determine all cyclic extensions $L/\mathbb{Q}$ of degree a power of a prime ℓ whose class number (in either sense) is not divisible by ℓ, and we shall also make the preliminary steps in solving the same problem (in §5) for Abelian extensions.

Consider a commutative diagram of groups and homomorphisms

$$\begin{array}{ccccccccc} 1 & \longrightarrow & \Delta & \longrightarrow & \Sigma & \longrightarrow & \Gamma & \longrightarrow & 1 \\ & & \downarrow 1 & & \downarrow & & \downarrow & & \\ 1 & \longrightarrow & \Delta & \longrightarrow & \Sigma' & \longrightarrow & \Gamma^{ab} & \longrightarrow & 1 \end{array} \tag{2.1}$$

where the lower row is given, with Σ' Abelian, and the right-hand column being the natural quotient map. We then call the top row the group extension of Γ by Δ <u>lifted from an Abelian extension</u>. (Of course $\Sigma' \cong \Sigma^{ab}$). Given an arbitrary group extension

$$1 \to \Delta_1 \to \Sigma_1 \to \Gamma \to 1, \tag{2.2}$$

we derive a maximal quotient extension, lifted in the above sense, and given by

$$1 \to \Delta_1/(\Delta_1 \cap (\Sigma_1,\Sigma_1)) \to \Sigma_1/(\Delta_1 \cap (\Sigma_1,\Sigma_1)) \to \Gamma \to 1. \tag{2.2a}$$

Now let (2.2) be an extension of Galois groups associated with a tower

$K \subset L \subset M$ of Galois extensions of fields. Thus $\Delta_1 = \mathrm{Gal}(M/L)$, $\Sigma_1 = \mathrm{Gal}(M/K)$, $\Gamma = \mathrm{Gal}(L/K)$. Then the fixed field E of (Σ_1,Σ_1) is the maximal Abelian extension of K in M, and we have the following diagram of fields and corresponding Galois groups.

(2.3)

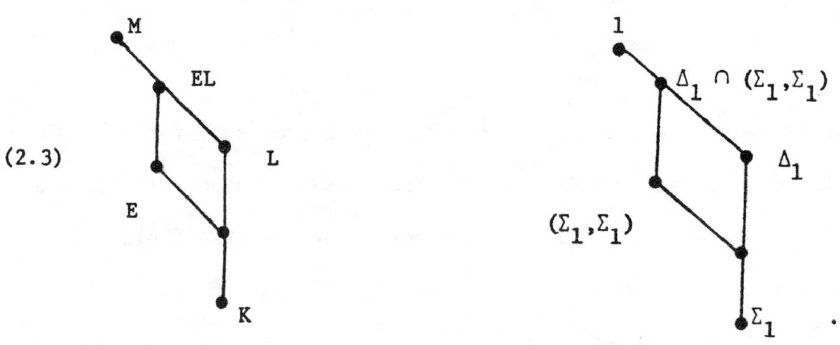

From now on assume that the fields in (2.3) are number fields. We shall need two results on the associated class groups.

2.1 Lemma. $N_{EL/L}(C(EL)) = N_{L/K}^{-1}(N_{E/K}(C(E)))$, *where* $N_{L/K}^{-1}$ *denotes the inverse image*.

Proof. From (1.6), (1.9) we get a commutative diagram with exact rows

$$\begin{array}{ccccccccc} 1 & \longrightarrow & N_{EL/L}(C(EL)) & \longrightarrow & C(L) & \longrightarrow & \mathrm{Gal}(EL/L) & \longrightarrow & 1 \\ & & \big\downarrow & & \big\downarrow N_{L/K} & & \big\downarrow & & \\ 1 & \longrightarrow & N_{E/K}(C(E)) & \longrightarrow & C(K) & \longrightarrow & \mathrm{Gal}(E/K) & \longrightarrow & 1. \end{array}$$

The right-hand column is injective, and with a bit of diagram chasing this gives the result.

2.2 Lemma. *With* M, L, K *as above and* M/L *Abelian, we have*

(a) $$N_{M/L}(C(M)) \supset \mathrm{Ker}\, N_{L/K},$$

if, and only if,

$$M = EL, \text{ with } E/K \text{ Abelian.}$$

Proof. Suppose (a) to hold. Let E be again the maximal Abelian extension of K in M. Then $N_{L/K}(N_{EL/L}(C(EL))) = N_{EL/K}(C(EL)) = N_{E/K}(C(E)) = N_{M/K}(C(M)) = N_{L/K}(N_{M/L}(C(M)))$. By (a) this implies that

$$N_{M/L}(C(M)) \supset N_{EL/L}(C(EL)), \quad \text{i.e.,} \quad M \subset EL,$$

hence $M = EL$.

The converse is immediate from Lemma 2.1.

If in (2.3) M is the absolute class field of L (the narrow absolute class field of L, respectively,) then EL is called the genus field (the narrow genus field, respectively) and denoted by $G(L/K)$ (by $G_+(L/K)$, respectively). Alternatively we may define $G(L/K)$, and $G_+(L/K)$, respectively, as the maximal extension of L, which is non-ramified, or non-ramified at all finite primes, respectively, and is the compositum of L and some Abelian extension of K.

We shall continue the discussion in terms of the genus field $G(L/K)$, with only occasional references to the narrow case, i.e., to $G_+(L/K)$, where of course analogous definitions and results apply.

The genus group $\mathfrak{G}(L/K)$ is defined as the class group associated with the genus field, i.e., by

$$\mathfrak{G}(L/K) = C(L)/N_{G/L}(C(G)), \quad G = G(L/K). \tag{2.4}$$

As G is a subfield of the absolute class field of L, $\mathfrak{G}(L/K)$ is a quotient of $\mathfrak{Cl}(L)$, and the principal genus is defined by

$$\mathfrak{H}(L/K) = \mathrm{Ker}[\mathfrak{Cl}(L) \to \mathfrak{G}(L/K)]. \tag{2.5}$$

It is easy to describe these groups more explicitly.

2.3 Lemma. *The inverse image in* $J(L)$ *of* $N_{G/L}(C(G))$ *is* $U(L)\cdot(N^{-1}_{L/K}(K^\times))$, *and so*

$$\mathfrak{G}(L/K) \cong J(L)/(U(L)\cdot(N^{-1}_{L/K}(K^\times))).$$

Proof. We consider the inverse images H_M in $J(L)$ of norm groups $N_{M/L}(C(M))$, with M/L Abelian. By Lemma 2.2, $H_M \supset N^{-1}_{L/K}(K^\times)$ precisely if M is composed from L and some Abelian extension of K.

$H_M \supset U(L)$ precisely if M is non ramified over L. But $U(L)\cdot N^{-1}_{L/K}(K^\times)$ is such a group H_M, hence it is H_G.

Now we can give the description of the genus group in terms of the base field K.

2.4 Proposition. The norm $N_{L/K}$ give rise to an isomorphism

$$\mathfrak{G}(L/K) \cong N_{L/K}(J(L))\cdot K^\times/N_{L/K}(U(L))\cdot K^\times.$$

Hence also

$$\mathfrak{G}(L/K) \cong N_{L/K}(J(L))/N_{L/K}(U(L))[N_{L/K}(J(L)) \cap K^\times].$$

This is an immediate consequence of Lemma 2.3.

Replacing $G(L/K)$ by $G_+(L/K)$ we get the narrow genus group $\mathfrak{G}_+(L/K)$, and replacing $\mathfrak{Cl}(L)$ by $\mathfrak{Cl}_+(L)$ in (2.5) we get the narrow principal genus $\mathfrak{H}_+(L/K)$. One then gets analogues of 2.3 and 2.4, with $U(L)$ replaced by $U(L)_+$.

Remarks. 1. Another way of expressing the result in the last proposition is by the equation

$$N_{G/K}(C(G)) = N_{L/K}(U(L))\cdot K^\times/K^\times, \text{ for } G = G(L/K).$$

2. It is immediate from the definition that $G(L/K)$ is its own genus field.

3. If $K^{(1)}$ is the absolute class field of K then of course $K^{(1)}L \subset G(L/K)$. In fact the inverse image in $J(L)$ of $N_{K^{(1)}L/L}(C(K^{(1)}L))$ is $N^{-1}_{L/K}(U(K)\cdot K^\times)$. Comparison with Lemma 2.3 gives a measure of how far $G(L/K)$ differs from $K^{(1)}L$. We shall not go onto this in detail, as we are mainly interested in the case when $K^{(1)} = K$, e.g., $K = \mathbb{Q}$.

The genus group behaves well with respect to "change of field". We only need one result.

2.5 Lemma. Let $L \supset M \supset K$ be a tower of number fields, with L/K and M/K Galois. The inclusion map $J(M) \to J(L)$ gives rise to a homomorphism

$$j_{L/M} : \mathfrak{G}(M/K) \to \mathfrak{G}(L/K).$$

and the norm map to a homomorphism

$$N_{L/M} : \mathfrak{G}(L/K) \to \mathfrak{G}(M/K).$$

Proof. Immediate from Lemma 2.3.

All these remarks and results have analogues in the narrow case.

For computations it is often more useful to look at the duals $\check{\mathfrak{G}}(L/K)$ and $\check{\mathfrak{G}}_+(L/K)$ and this approach also gives a better insight into certain structural properties. Recall first a local result:

2.6 Lemma. Let E/F be an extension of local fields. Then the conorm $\check{N}_{E/F}$ is surjective on non ramified characters.

Indeed in the non Archimedean case the conorm on non ramified characters is effectively the dual of multiplication in $\mathbb{Z}$ by the residue class degree. In the Archimedean case there is nothing to prove.

The non-triviality of genus theory is due to the fact that the global analogue to the last lemma is false. Clearly

$$\check{\mathfrak{G}}(L/K) = \mathfrak{G}(L/K)^{\dagger} = [\phi \in \operatorname{Im} \check{N}_{L/K} | \phi \text{ non ramified}]. \tag{2.6}$$

It is then natural to look at the group

$$\Phi^{*}(L/K) = [\phi \in C(K)^{\dagger} | \check{N}_{L/K}\phi \text{ is non ramified}]. \tag{2.7}$$

We clearly get an exact sequence

$$1 \to \Phi(L/K) \to \Phi^{*}(L/K) \to \check{\mathfrak{G}}(L/K) \to 1. \tag{2.8}$$

Now we have

<u>2.7 Proposition</u>. (i) $\Phi^*(L/K) = \Phi(G(L/K)/K)$.

(ii) $\Phi^*(L/K)$ <u>is the subgroup of</u> $C(K)^\dagger$ <u>of characters</u> ϕ <u>with the following property</u>: <u>For each prime divisor</u> $\mathfrak{p}$ <u>of</u> K,

$$\phi_{\mathfrak{p}} = \phi_{\mathfrak{p}}^{(1)}\phi_{\mathfrak{p}}^{(2)},\ \phi_{\mathfrak{p}}^{(1)}\ \underline{\text{non ramified}},\ \phi_{\mathfrak{p}}^{(2)} \in \Phi(L_{\mathfrak{P}}/K_{\mathfrak{p}})\ \underline{\text{with}}\ \mathfrak{P}\ \underline{\text{a}}\ \underline{\text{prime divisor of}}\ L\ \underline{\text{above}}\ \mathfrak{p}. \tag{2.9}$$

<u>Proof</u>. (i) is a clear consequence of the definitions. For (ii) first observe that evidently characters ϕ, satisfying (2.9) for all $\mathfrak{p}$, will lie in $\Phi^*(L/K)$. Conversely if $\phi \in \Phi^*(L/K)$, then for all $\mathfrak{p}$, $\check{N}_{L_{\mathfrak{P}}/K_{\mathfrak{p}}}\phi_{\mathfrak{p}}$ is non ramified, $\mathfrak{P}$ as in the proposition. But by Lemma 2.6 this implies (2.9).

<u>Corollary</u>. <u>Suppose that</u> $\mathfrak{Cl}(K) = 1$. <u>Let</u> m <u>be the exponent of</u> $\Gamma = \mathrm{Gal}(L/K)$. <u>Then</u> $\mathfrak{G}(L/K)^m = 1$.

For, in (2.9) always $\phi_{\mathfrak{p}}^{(2)m} = 1$. Hence ϕ^m is non ramified. As $\mathfrak{Cl}(K) = 1$, $\phi^m = 1$. Thus $\mathfrak{G}(L/K)^m = 1$.

Next, we establish a lemma, which in particular will allow us to reduce the study of the genus group for Abelian extensions to that of extensions of prime power degree, provided that $\mathfrak{Cl}(K) = 1$. It is clear of course that there is a more general formulation of this lemma as well as of the above corollary dealing with the general case - however, for us this is of no particular interest.

<u>2.8 Lemma</u>. <u>Let</u> $\mathfrak{Cl}(K) = 1$. <u>Let</u> M_i/K $(i = 1,2)$ <u>be Galois extensions of coprime degrees</u> m_i $(i = 1,2)$ <u>and let</u> $L = M_1M_2$. <u>Then</u>

$$\mathfrak{G}(L/K) = \mathfrak{G}^{(1)}(L/K) \times \mathfrak{G}^{(2)}(L/K)$$

<u>where</u> $\mathfrak{G}^{(i)}(L/K)^{m_i} = 1$ <u>and where the maps</u> j_{L/M_i}, N_{L/M_i} <u>of Lemma 2.5 give rise to isomorphisms</u>

$$\mathfrak{G}(M_i/K) \rightleftarrows \mathfrak{G}^{(i)}(L/K).$$

<u>Proof</u>. The direct product expression for $\mathfrak{G}(L/K)$ is a consequence of the above corollary, and by the same corollary the maps $x \longrightarrow x^{m_2}$ are automorphisms

of $\mathfrak{G}(M_1/K)$, as well as, of course, of $\mathfrak{G}^{(1)}(L/K)$. But the composite map $N_{L/M_1} \circ j_{L/M_1}$ is clearly just $x \longmapsto x^{m_2}$, and as $Gal(L/M_1)$ acts trivially on $\mathfrak{G}(L/K)$, the same is true for $j_{L/M_1} \circ N_{L/M_1}$.

Lemmas 2.5 and 2.8 remain valid for $\mathfrak{G}_+$, on replacing the hypothesis on the absolute class group in the appropriate place by $\mathfrak{Cl}_+(K) = 1$. To describe the dual of $\mathfrak{G}_+$ one has to replace $\Phi^*(L/K)$ by

$$\Phi_+^*(L/K) = [\phi \in C(K)^\dagger | \check{N}_{L/K}\phi \text{ is non ramified at all finite primes}]. \quad (2.7a)$$

Then Proposition 2.7 goes over, with (2.9) restricted to all finite primes, and so does the Corollary, given that $\mathfrak{Cl}_+(K) = 1$.

We now compare the two genus fields.

<u>2.9 Proposition</u>. $G(L/K)$ <u>is always the maximal subfield of</u> $G_+(L/K)$, <u>which is real above all real primes of</u> L. <u>In particular if</u> L <u>is totally imaginary, then</u> $G(L/K) = G_+(L/K)$. <u>The same is true if</u> $\mathfrak{Cl}(K) = 1$ <u>and</u> $[L:K]$ <u>is odd</u>.

<u>Proof</u>. The first statement is am immediate consequence of the definitions, the second statement an obvious corollary -- in fact, if L is totally imaginary then $U(L) = U(L)_+$.

Next if $\check{\mathfrak{G}}(L/K)$, and $\check{\mathfrak{G}}_+(L/K)$ are viewed as subgroups of $C(L)^\dagger$, then, as $U(L)^2 \subset U(L)_+ \subset U(L)$, also

$$\check{\mathfrak{G}}_+(L/K)^2 \subset \check{\mathfrak{G}}(L/K) \subset \check{\mathfrak{G}}_+(L/K).$$

But by the Corollary to Proposition 2.7, also $\check{\mathfrak{G}}_+(L/K)^m \subset \check{\mathfrak{G}}(L/K)$, for some odd integer m. Therefore, indeed $\check{\mathfrak{G}}_+(L/K) = \check{\mathfrak{G}}(L/K)$.

We now come to the special role of the genus group for cyclic extensions. In the next proposition we use the fact that if Γ is cyclic of prime order ℓ, and if X is a $\mathbb{Z}\,\Gamma$-module with $X^\Gamma = 0$, (X^Γ the set of fixed point), then X may be viewed as a module over the ring $\mathbb{Z}[\ell]$ of integers in the field of ℓ-th roots of unity over $\mathbb{Q}$. We again leave the "narrow" analogue to the reader.

2.10 Proposition. Let L/K be cyclic, with Gal(L/K) generated by an element σ. Then

(i) $\mathfrak{H}(L/K) = \mathfrak{Cl}(L)^{1-\sigma}$, i.e., $\mathfrak{G}(L/K) = \mathfrak{Cl}(L)/\mathfrak{Cl}(L)^{1-\sigma}$.

If moreover $[L : K]$ is a power of a prime ℓ and A_ℓ denotes the ℓ-sylow group of a group A, then

(ii) $\mathfrak{G}(L/K)_\ell = 1$ if and only if $\mathfrak{Cl}(L)_\ell = 1$.

If in addition $[L : K] = \ell$ and $\mathfrak{Cl}(K) = 1$ then

(iii) $\dim_{\mathbb{F}_\ell}(\mathfrak{G}(L/K))_\ell$ = minimal number of generators of $\mathfrak{Cl}(L)_\ell$ over $\mathbb{Z}[\ell]$.

Note in connection with (ii) that, by the Corollary to Proposition 2.7, $\mathfrak{G}(L/K)_\ell = \mathfrak{G}(L/K)$ if $\mathfrak{Cl}(K) = 1$.

Proof of 2.10. Let $\Omega = \mathrm{Gal}(L^{(1)}/K)$, $L^{(1)}$ the absolute class field of L and let L be the fixed field of the subgroup Δ. The reciprocity isomorphism $\mathfrak{Cl}(L) \cong \Delta$ maps $\mathfrak{Cl}(L)^{1-\sigma}$ onto the commutator group (Δ,Ω) and $\mathfrak{H}$ onto (Ω,Ω). As Ω/Δ is cyclic we know that $(\Delta,\Omega) = (\Omega,\Omega)$. This gives (i).

Next assume that $[L : K]$ is a power of ℓ. Let L' be the maximal subextension of $L^{(1)}$ whose degree is a power of ℓ, let $\Omega' = \mathrm{Gal}(L'/K)$, $\Delta' = \mathrm{Gal}(L'/L)$. If $\mathfrak{G}(L/K)_\ell = 1$, then $\Delta' = (\Omega',\Omega')$. As Ω' is an ℓ-group, and Ω'/Δ' is cyclic, this implies that $\Delta' = 1$, and so $\mathfrak{Cl}(L)_\ell = 1$. The converse is trivial.

Finally, under the hypothesis for (iii) the sum $\Sigma\gamma$ over the elements of Gal(L/K) annihilates $\mathfrak{Cl}(L)_\ell$, which thus becomes a finite $\mathbb{Z}_\ell[\ell]$-module. But, by part (i), $\mathfrak{G}(L/K)_\ell = \mathfrak{Cl}(L)_\ell/\mathfrak{Cl}(L)_\ell^J$, J the radical of $\mathbb{Z}_\ell[\ell]$. This then immediately gives (iii) by standard module theory.

The classical principal genus theorem. We first prove a general result which will also be needed later.

2.11 Lemma. Let L/K be a Galois extension of number fields with Galois group Γ. Let I(L) be the group of fractional ideals in L. Then $\hat{H}^{-1}(\Gamma,I(L)) = 1$. (Here $\hat{H}$ denotes Tate cohomology.)

Recall the definition of $\hat{H}^{-1}(\Gamma,X)$ for a Γ-module X. This is the quotient $(\mathrm{Ker}\ \Sigma\gamma_{|X})/XA(\Gamma)$, where $\Sigma\gamma_{|X}$ is the trace map $x \longmapsto \Sigma x\gamma$ on X, $A(\Gamma)$ is the augmentation ideal and so $XA(\Gamma)$ is the submodule generated by elements of form $x - x\gamma = x(1-\gamma)$.

<u>Proof of</u> 2.11. $I(L)$ is the direct sum of Γ-modules $I_{\mathfrak{p}}(L)$, with $\mathfrak{p}$ running over the prime ideals of the integer ring of K; $I_{\mathfrak{p}}(L)$ consist of those fractional ideals, which only involve prime divisors of $\mathfrak{p}$. We have to show that $\hat{H}^{-1}(\Gamma,I_{\mathfrak{p}}(L)) = 1$. But if Δ is the decomposition group of a prime divisor of $\mathfrak{p}$ in L, then as Γ-modules $I_{\mathfrak{p}}(L) \cong \mathbb{Z} \otimes_{\mathbb{Z}\Delta} \mathbb{Z}\Gamma$, hence $\hat{H}^{-1}(\Gamma,I_{\mathfrak{p}}(L)) \cong \hat{H}^{-1}(\Delta,\mathbb{Z}) = 1$.

To state the classical result, without having to introduce too many new definitions, let for each prime divisor $\mathfrak{p}$ of K, $\mathfrak{P}$ be a prime divisor in L above $\mathfrak{p}$.

<u>2.12 Proposition</u>. (The principal genus theorem). <u>Let</u> L/K <u>be a cyclic extension</u>, σ <u>a generator of</u> $\mathrm{Gal}(L/K)$. <u>The class of a fractional ideal</u> $\mathfrak{a}$ <u>of</u> L <u>will lie in the principal genus</u> $\mathfrak{H}(L/K)$ <u>if</u>, <u>and only if</u>, $N_{L/K}(\mathfrak{a}) = (a)$ <u>with</u> $a \in K^{\times}$ <u>and</u> $a_{\mathfrak{p}} \in N_{L_{\mathfrak{P}}/K_{\mathfrak{p}}}(L_{\mathfrak{P}}^{\times})$ <u>for all ramified prime divisors</u> $\mathfrak{p}$, <u>including the infinite ones</u>.

<u>Remark</u>. This proposition has no direct analogue in the narrow case.

<u>Proof of</u> 2.12. If the class of $\mathfrak{a}$ lies in $\mathfrak{H}(L/K)$, then, by (2.10(i)), $\mathfrak{a} = \mathfrak{b}^{1-\sigma}(c)$, and so $N_{L/K}(\mathfrak{a}) = (a)$ with $a = N_{L/K}(c)$. This immediately gives the stated condition. Conversely the equation $N_{L/K}(\mathfrak{a}) = (a)$, with $a \in K^{\times}$, implies certainly that $a \in N_{L_{\mathfrak{P}}/K_{\mathfrak{p}}}(L_{\mathfrak{P}}^{\times})$ whenever $\mathfrak{p}$ is non ramified in L. Thus the stated condition implies that $a \in N_{L_{\mathfrak{P}}/K_{\mathfrak{p}}}(L_{\mathfrak{P}}^{\times})$, for all $\mathfrak{p}$. In other words

$$a \in N_{L/K}(J(L)),$$

i.e., the class of a in $\hat{H}^{\circ}(\Gamma,L^{\times})$ lies in

$$\mathrm{Ker}[\hat{H}^{\circ}(\Gamma,L^{\times}) \to H^{\circ}(\Gamma,J(L))] = \mathrm{Im}[\hat{H}^{-1}(\Gamma,C(L)) \to \hat{H}^{\circ}(\Gamma,L^{\times})],$$

using the cohomology sequence associated with the exact sequence

$$1 \to L^{\times} \to J(L) \to C(L) \to 1.$$

But as Γ is cyclic $\hat{H}^{-1}(\Gamma,C(L)) = \hat{H}^{1}(\Gamma,C(L)) = 1$. Thus in fact

$$\mathfrak{a} = N_{L/K}(c),$$

i.e., $N_{L/K}(\mathfrak{a}(c^{-1})) = (1)$, whence by Lemma 2.11, $\mathfrak{a} = (c)\mathfrak{b}^{1-\sigma}$. This is what we had to show.

Rational basefield. From now on, for the remainder of this section we take $K = \mathbb{Q}$. By (1.21) we have a decomposition

$$J(\mathbb{Q}) = U_{fin}(\mathbb{Q}) \times \mathbb{Q}^{*} \times \mathbb{R}_{+}^{*} \quad (\mathbb{R}_{+}^{*} \text{ the positive reals}). \tag{2.10}$$

The projection $J(\mathbb{Q}) \to U_{fin}(\mathbb{Q})$ will be denoted by P. If M is an Abelian number field we shall write

$$\mathrm{Im}[\mathrm{Gal}(\mathbb{Q}^{ab}/M) \to U_{fin}(\mathbb{Q})] = V(M/\mathbb{Q}),$$

the map from the Galois group being that of (1.23), (1.24), i.e., given by the action on roots of unity.

2.13 Proposition. *Let $L/\mathbb{Q}$ be a Galois extension.*

(i) $P(N_{L/\mathbb{Q}}(J(L))\cdot\mathbb{Q}^{*}) = V(L'/\mathbb{Q}),$

L' *the maximal Abelian subfield of* L.

(ii) $P(N_{L/\mathbb{Q}}(U(L)_{+})\cdot\mathbb{Q}^{*}) = N_{L/\mathbb{Q}}(U_{fin}(L)),$

and if L *is real*

(iii) $P(N_{L/\mathbb{Q}}(U(L))\cdot\mathbb{Q}^{*}) = N_{L/\mathbb{Q}}(U_{fin}(L))\,(\pm 1)_{fin},$

where $(\pm 1)_{fin}$ *is the image of* ± 1 *in* $U_{fin}(\mathbb{Q})$.

P *gives rise to isomorphisms*

$$\mathfrak{C}_{+}(L/\mathbb{Q}) \cong V(L'/\mathbb{Q})/N_{L/\mathbb{Q}}(U_{fin}(L)),$$

<u>and if</u> L <u>is real</u>,

$$\mathfrak{G}(L/\mathbb{Q}) \simeq V(L'/\mathbb{Q})/N_{L/\mathbb{Q}}(U_{fin}(L))(\pm 1)_{fin}.$$

<u>Proof</u>. (i) is immediate by (1.7) and the properties of (1.22).

For (ii) we only have to note that the product decomposition (2.10) is compatible with the decomposition

$$(N_{L/\mathbb{Q}}(U(L)_+)\cdot\mathbb{Q}^* = N_{L/\mathbb{Q}}(U_{fin}(L)) \times \mathbb{Q}^* \times \mathbb{R}_+^*.$$

If L is real, then the infinite component of $N_{L/\mathbb{Q}}(U(L))$ is $(\pm 1)_\infty\cdot \mathbb{R}_+^*$, where $(\pm 1)_\infty$ is the image of ± 1 in $\mathbb{R}^*$. But $(-1)_\infty = (-1)(-1)_{fin} \in U_{fin}(\mathbb{Q}) \times \mathbb{Q}^*$. Therefore

$$N_{L/\mathbb{Q}}(U(L))\cdot\mathbb{Q}^* = (N_{L/\mathbb{Q}}(U_{fin}(L))\cdot(\pm 1)_{fin}) \times \mathbb{Q}^* \times \mathbb{R}_+^*,$$

and this gives (iii).

The isomorphisms are a consequence of (i)-(iii) and the fact that $\mathrm{Ker}\ P \subset N_{L/\mathbb{Q}}(U(L)_+)\cdot\mathbb{Q}^*$.

<u>Corollary 1</u>. <u>Each of the following three statements implies the others</u>.

(i) $\mathfrak{G}_+(L/\mathbb{Q}) = \mathfrak{G}(L/\mathbb{Q})$

(ii) L <u>and</u> $G_+(L/\mathbb{Q})$ <u>are both complex</u>, <u>or are both real</u>.

(iii) L <u>is complex</u>, <u>or</u> L <u>is real and</u> $-1 \in N_{L/\mathbb{Q}}(J(L))$.

This follows from Proposition 2.13 in conjunction with Proposition 2.9. Observe also that if $-1 \in N_{L/\mathbb{Q}}(J(L))$ then actually $-1 \in N_{L/\mathbb{Q}}(U(L))$, and conversely if $(-1)_{fin} \in N_{L/\mathbb{Q}}(U_{fin}(L))$ and L is real, then $-1 \in N_{L/\mathbb{Q}}(J(L))$.

Let now S be a finite set of finite rational primes, including those ramified in L. Put

$$U_S = \prod_{p\in S} U(\mathbb{Q}_p), \quad U^S = \prod_{\substack{p\notin S\\ p\neq\infty}} U(\mathbb{Q}_p).$$

Then

$$U_{fin}(\mathbb{Q}) = U_S \times U^S$$

and

$$U^S \subset N_{L/\mathbb{Q}}(U(L)_+).$$

Let P_S be the compositum of P and the projection $U_{fin}(\mathbb{Q}) \longrightarrow U_S$. Denote by $V(L'/\mathbb{Q})_S$ the image of $V(L'/\mathbb{Q})$ under this projection. Then we have

Corollary 2. *P_S gives rise to isomorphisms*

$$\mathfrak{G}_+(L/\mathbb{Q}) \cong V(L'/\mathbb{Q})_S/(\prod_{p\in S} N_{L_\mathfrak{p}/\mathbb{Q}_p}(U(L_\mathfrak{p}))),$$

and for L real,

$$\mathfrak{G}(L/\mathbb{Q}) = V(L'/\mathbb{Q})_S/(\prod_{p\in S} N_{L_\mathfrak{p}/\mathbb{Q}_p}(U(L_\mathfrak{p})))(\pm 1)_S.$$

Here $\mathfrak{p}$ is always a prime of L lying above p and $(\pm 1)_S$ is the image of (± 1) in U_S.

Example: Let ℓ be an odd prime, η_ℓ a primitive ℓ-th root of unity, m an ℓ-power free integer, $L = \mathbb{Q}(\eta_\ell, m^{1/\ell})$. Then

$$\mathfrak{G}(L/\mathbb{Q}) \cong \mathbb{F}_\ell^r, \quad r = \text{number of primes } p \equiv 1 (\text{mod } \ell) \text{ dividing } m. \tag{2.11}$$

Indeed $L' = \mathbb{Q}(\eta_\ell)$. S may be taken as the set consisting of ℓ and all primes dividing m. We then have

$$V(L'/\mathbb{Q})_S = U^{(1)}(\mathbb{Q}_\ell)\cdot \prod_{\substack{p\in S\\ p\neq \ell}} U(\mathbb{Q}_p),$$

where $U^{(1)}(\mathbb{Q}_\ell)$ is the group of units $\equiv 1$ (mod ℓ) in $\mathbb{Q}_\ell$. Also $U^{(1)}(\mathbb{Q}_\ell) = N_{L_\mathfrak{p}/\mathbb{Q}_\ell}(U(L_\mathfrak{p}))$, $\mathfrak{p}$ above ℓ. For, the maximal Abelian extension of $\mathbb{Q}_\ell$ in $L_\mathfrak{p}$ is the field of ℓ-th roots of unity over $\mathbb{Q}_\ell$. Next if $p \neq \ell$, $p|m$, $p \not\equiv 1 (\text{mod } \ell)$, then again the field of ℓ-th roots over $\mathbb{Q}_p$, which is non-ramified over $\mathbb{Q}_p$, is the maximal Abelian extension of $\mathbb{Q}_p$ in L. Thus $U(\mathbb{Q}_p) = N_{L_\mathfrak{p}/\mathbb{Q}_p}(U(L_\mathfrak{p}))$. Finally if $p|m$, $p \equiv 1 (\text{mod } \ell)$, then $L_\mathfrak{p}$ is Abelian of degree ℓ over $\mathbb{Q}_p$ and totally ramified. Therefore $N_{L_\mathfrak{p}/\mathbb{Q}_p}(U(L_\mathfrak{p})) = U(\mathbb{Q}_p)^\ell$, which is of index ℓ in $U(\mathbb{Q}_p)$. Collecting all the information we now get (2.11).

We shall now - for $L/\mathbb{Q}$ Abelian - turn again to a description via characters. As we have seen (cf. (1.20)), we may and shall identify $C(\mathbb{Q})^{\dagger}$ and $U_{fin}(Q)^{\dagger}$. Some care is however needed as the notion of a local component means something quite different in the two cases. If $\phi \in C(\mathbb{Q})^{\dagger}$ we shall, as hitherto, denote its p-component by ϕ_p; thus ϕ_p is the compositum

$$\mathbb{Q}_p^* \to J(\mathbb{Q}) \to C(\mathbb{Q}) \overset{\phi}{\to} \mathbb{C}^*.$$

On the other hand, given $\phi \in U_{fin}(\mathbb{Q})^{\dagger}$ and a finite rational prime p, there is a unique character $\phi_{(p)}$ of $U_{fin}(\mathbb{Q})$ with the property that

$$\phi_{(p)}|U(\mathbb{Q}_p) = \phi|U(\mathbb{Q}_p),$$
$$\phi_{(p)}|U(\mathbb{Q}_q) = 1 \quad \text{if} \quad q \neq p.$$

Thus ϕ has only finitely many "components" $\phi_{(p)}$, which are non-trivial; these lie exactly at the primes at which ϕ is ramified. If Φ is a finite subgroup of $C(\mathbb{Q})^{\dagger}$ then the $\phi_{(p)}$, for all $\phi \in \Phi$ form again a finite subgroup of $C(\mathbb{Q})^{\dagger}$, which we shall denote by $\Phi_{(p)}$.

<u>2.14 Theorem</u>. <u>Let</u> L <u>be an Abelian extension of</u> $\mathbb{Q}$. <u>Then</u>

$$\Phi_+^*(L/\mathbb{Q}) = \prod_p \Phi(L/\mathbb{Q})_{(p)} \tag{2.12}$$

(<u>product over all finite primes</u>, <u>or all finite primes ramified in</u> L).

<u>If</u> L <u>is real then</u> $\Phi^*(L/\mathbb{Q})$ <u>is the subgroup of</u> $\Phi_+^*(L/\mathbb{Q})$ <u>of characters</u> ϕ, <u>with</u> $\phi((-1)_{fin}) = 1$.

(Of course, if L is complex then $\Phi_+^*(L/\mathbb{Q}) = \Phi^*(L/\mathbb{Q})$).

<u>Proof</u>. We prove (2.12) by going back to the narrow analogue of Proposition 2.7.

As $L/\mathbb{Q}$ is Abelian the group $\Phi(L/\mathbb{Q})_p$ of p-components - in the ordinary sense - is exactly the group $\Phi(L_{\mathfrak{p}}/\mathbb{Q}_p)$. (Use the commutativity of (1.10a)). By (2.9), $\Phi(L/\mathbb{Q})_{(p)}$ will thus be contained in $\Phi_+^*(L/\mathbb{Q})$. Conversely every $\phi \in C(\mathbb{Q})^{\dagger}$ is the product $\phi = \prod_p \phi_{(p)}$. If $\phi \in \Phi_+^*(L/\mathbb{Q})$, then $\phi_{(p)} \in \Phi_+^*(L/\mathbb{Q})$

as one easily verifies by (2.9). This implies in turn that actually $\phi_{(p)} \in \Phi(L/\mathbb{Q})_{(p)}$. This then gives (2.12).

Now suppose that L is real. Let $\phi \in \Phi_+^*(L/\mathbb{Q})$. By (2.9), $\phi \in \Phi^*(L/K)$ precisely if $\phi_\infty = 1$, i.e., if $\phi((-1)_\infty) = 1$. But $\phi((-1)_\infty) = \phi((-1)_{fin})$. Hence the result.

<u>Corollary</u>. <u>If</u> $L/\mathbb{Q}$ <u>is Abelian then</u>

$$\check{\mathfrak{G}}_+(L/\mathbb{Q}) \cong \prod_p \Phi(L/\mathbb{Q})_{(p)}/\Phi(L/\mathbb{Q}).$$

<u>Quadratic fields</u>. If $L/\mathbb{Q}$ is quadratic then Kummer theory becomes available. There is a bijection

$$d \leftrightarrow \phi$$

between the discriminants d of quadratic fields L and the quadratic characters, i.e., the characters ϕ in $C(\mathbb{Q})^\dagger$ of order 2: ϕ generates $\Phi(\mathbb{Q}(\sqrt{d})/\mathbb{Q})$. Each discriminant d is uniquely the product of prime power discriminants d_p. For p an odd prime there is one such discriminant, namely $(-1/p)p$. For $p = 2$ there are three possible such discriminants, -4, 8, and -8. Let then S be the set of prime factors of d. Then $d = \prod_S d_p$, and for $L = \mathbb{Q}(\sqrt{d})$ we have

$$G_+(L/\mathbb{Q}) = \mathbb{Q}(\sqrt{d_p}, \text{ all } p \in S).$$

We shall now give a criterion for an Abelian field L to be its own genus field over $\mathbb{Q}$ (in either sense). This is an important property, specifically for the study of class groups.

Note that, as $\mathrm{Gal}(L/\mathbb{Q}) = \Gamma$ is now Abelian, the inertia group $\Gamma_{p,0}$ of a prime divisor $\mathfrak{P}$ of p in L will indeed only depend on p.

<u>2.15 Theorem</u>. <u>Let</u> $L/\mathbb{Q}$ <u>be Abelian</u>, <u>denote by</u> S <u>the set of finite rational primes</u>, <u>ramified in</u> L.

(A) <u>Then the following four statements are equivalent</u>.

(i) $L = G_+(L/\mathbb{Q})$ <u>is its own narrow genus field</u>.

(ii) $\Phi(L/\mathbb{Q}) = \prod_S \Phi(L/\mathbb{Q})_{(p)}$

(iii) $\mathrm{Gal}(L/\mathbb{Q})$ _is the direct sum of its inertia groups_ $\Gamma_{p,0}$, p _running through_ S.

(iv) L _is the composite of fields with prime power discriminant_.

(B) $L = G(L/\mathbb{Q})$ _precisely in the following two cases_:

(a) $L = G_+(L/\mathbb{Q})$

(b) L _is the maximal real subfield of a complex Abelian field_ M, _with_ $M = G_+(M/\mathbb{Q})$, _and_ M/L _is non ramified at all finite primes_.

Proof. (A) The equivalence of (i) and (ii) follows from the Corollary to Theorem 2.14. For, L is its own narrow genus field precisely when $\mathfrak{G}_+(L/\mathbb{Q}) = 1$, i.e., $\check{\mathfrak{G}}_+(L/\mathbb{Q}) = 1$.

Next, we may view characters as characters of Galois groups, the map $\phi \longmapsto \phi_{(p)}$ being viewed as restriction to inertia subgroups. Thus the dual of the embedding

$$\Phi(L/\mathbb{Q}) \to \prod_S \Phi(L/\mathbb{Q})_{(p)}$$

is the surjection

$$\coprod_S \Gamma_{p,0} \to \mathrm{Gal}(L/\mathbb{Q}),$$

and if one is an isomorphism so is the other. This establishes the equivalence of (ii) and (iii).

Finally the equivalence of (iii) and (iv) follows from a simple application of Galois Theory.

(B) If $L = G_+(L/K)$, then as $L \subset G(L/K) \subset G_+(L/K)$ also $L = G(L/K)$.

Now suppose that $L = G(L/K) \neq G_+(L/K)$. By Corollary 1 to Proposition 2.13, L is real, but $G_+(L/K)$ is not, whence by Proposition 2.9, $L = G(L/K)$ is the maximal real subfield of $G_+(L/K)$, and of course, by definition, $G_+(L/K)/L$ is non ramified at all finite primes. Conversely if M satisfies the conditions in (B) (b), then as $L \subset M$ also $G_+(L/\mathbb{Q}) \subset G_+(M/\mathbb{Q}) = M$. But as M/L

is non ramified at all finite primes and $M/\mathbb{Q}$ is Abelian, also $M \subset G_+(L/\mathbb{Q})$. Hence indeed $M = G_+(L/\mathbb{Q})$. By Proposition 2.9, $L = G(L/\mathbb{Q})$.

We finally look at cyclic extensions $L/\mathbb{Q}$. This was originally the main area of application of genus theory. We shall in fact restrict ourselves to cyclic extensions of prime power degree. By Lemma 2.8 this does not entail any loss of generality, while on the other hand the results look much neater in this case. Moreover, because of Proposition 2.10, there is a special interest in prime power degree cyclic extensions.

<u>2.16 Theorem</u>. <u>Let</u> $L/\mathbb{Q}$ <u>be a cyclic extension of degree</u> ℓ^n, ℓ <u>a prime</u>. <u>Let</u> $p_i (i = 1,\dots,t)$ <u>be the finite rational primes</u>, <u>ramified in</u> L, <u>of ramification degrees</u> ℓ^{n_i}, $n_1 \geq n_2 \geq \cdots \geq n_t \geq 1$. <u>Then</u> $\mathfrak{G}_+(L/\mathbb{Q})$ <u>is Abelian of type</u> $(\ell^{n_2},\dots,\ell^{n_t})$, <u>and in particular the minimal number of generators is</u> $t - 1$.

The order of $\mathfrak{Cl}_+(L)$, the so-called <u>narrow class number</u> of L will be denoted by $h_+(L)$.

From the Theorem and Proposition 2.10 we conclude

<u>Corollary 1</u>. $\mathfrak{Cl}_+(L)_\ell = 1$, <u>i.e.</u>, $\ell \nmid h_+(L)$ <u>if, and only if</u>, $t = 1$.

Also

<u>Corollary 2</u>. $\ell^{n_2+\cdots+n_t} \mid h_+(L)$.

<u>Proof of Theorem 2.16</u>. As $L/\mathbb{Q}$ is totally ramified we must have $n = n_1$. $\Phi(L/\mathbb{Q})$ is cyclic of order ℓ^n. For each i, $\Phi(L/\mathbb{Q})_{(p_i)}$ is cyclic of order ℓ^{n_i}, the ramification degree of p_i. The theorem now follows from the Corollary to Theorem 2.14.

The last theorem still leaves open the question what precisely $\mathfrak{G}(L/\mathbb{Q})$ looks like, in the case when $\ell = 2$ and L is real. We have a surjection $\mathfrak{G}_+(L/\mathbb{Q}) \to \mathfrak{G}(L/\mathbb{Q})$ whose kernel is of order 1 or 2, and it is the former precisely when -1 is norm residue, i.e., $-1 \in N_{L/\mathbb{Q}}(J(L))$, e.g. by Proposition 2.13. One can in fact determine $\mathfrak{G}(L/\mathbb{Q})$ again quite explicitly. We shall however simply content ourselves with obtaining an analogue to Corollary 1 of the last theorem. Let $h(L) = \text{order } \mathfrak{Cl}(L)$ (the class number).

2.17 Theorem. *Let* $L/\mathbb{Q}$ *be a cyclic extension of degree* $2^n (n \geq 1)$. *Then* $h(L)$ *is odd if, and only if one of the following conditions holds*

(a) *Precisely one finite rational prime is ramified in* L

(b) L *is real and precisely two rational primes* p_1 *and* p_2 *are ramified in* L *of ramification degrees* 2^n, *and* 2, *respectively and furthermore either* $p_2 \equiv -1 \pmod 4$, *or* $p_2 = 2$ *and* $\phi((-1)_2) = -1$, *where* ϕ *is the generator of* $\Phi(L/\mathbb{Q})$.

Remark. The last conditions says that ϕ_2 is one of the two quadratic characters on $U(\mathbb{Q}_2)$ with value -1 on $(-1)_2$.

Proof. By Proposition 2.10, $h(L)$ is odd if, and only if $[G(L/\mathbb{Q}) : L] = 1$. Condition (a) gives us, by Theorem 2.16, the case when actually $[G_+(L/\mathbb{Q}) : L] = 1$. So we now assume that $G_+(L/\mathbb{Q}) \neq L$. By Theorem 2.15, $L = G(L/\mathbb{Q})$ now implies that L is real and that $[G_+(L/\mathbb{Q}) : L] = 2$, i.e., $\mathfrak{G}_+(L/\mathbb{Q})$ is of order 2. By Theorem 2.16, this in turn implies that precisely two primes p_1, p_2 are ramified, with the ramification degrees as given. Assume that this is so. Then by Corollary 1 of Proposition 2.13, we shall get $G(L/\mathbb{Q}) \neq G_+(L/\mathbb{Q})$, i.e., $G(L/\mathbb{Q}) = L$, precisely if $-1 \notin N_{L/\mathbb{Q}}(J(L))$. If $\phi((-1)_{p_2}) = -1$ this last inequality will hold. Now assume that $\phi((-1)_{p_2}) = 1$. As L is real also $\phi((-1)_\infty) = 1$, and of course for all $p \neq p_1, p_2$, $\phi((-1)_p) = 1$. By the product formula $\phi((-1)_{p_1}) = 1$. Thus $-1 \in N_{L/\mathbb{Q}}(J(L))$. We conclude: Suppose that we are not in case (a). Then $L = G(L/\mathbb{Q})$ if, and only if, precisely two rational primes p_1 and p_2 are ramified, of ramification degrees 2^n and 2, and $\phi((-1)_{p_2}) = -1$. This is exactly what was asserted in the theorem.

Corollary. The quadratic fields with odd class number are the following

$L = \mathbb{Q}(\sqrt{p})$, p *any prime*,

$L = \mathbb{Q}(\sqrt{-p})$, p *any prime* $\equiv -1 \pmod 4$

or $p = 2$

$L = \mathbb{Q}(\sqrt{-1})$,

$L = \mathbb{Q}(\sqrt{p_1 p_2})$ p_1, p_2 *distinct primes*,

$p_1 \equiv -1 \pmod 4$, $p_2 = 2$ *or* $p_2 \equiv -1 \pmod 4$.

§3. Central Extensions

The study of central extensions of number fields leads to a number of arithmetic applications, some of which will be discussed in detail subsequently. Our basic tool - and in this we differ from some other accounts - is the cohomology group $\hat{H}^{-1}(\Gamma, C(L))$ for some Galois extension L/K of number fields with Galois group Γ. There is also a special theorem, valid over $\mathbb{Q}$ but not in general over any number field K as base field, which leads to important consequences.

We mention some of the applications. 1) We get some fairly deep information about the maximal ℓ-extension of $\mathbb{Q}$ (ℓ any prime including 2), and about the maximal ℓ-extension with ramification restricted to a finite set of primes - specifically in terms of their Galois groups and the inertia subgroups. In particular we shall also obtain a complete description of the Galois group of the maximal extension which is nilpotent of class two. 2) One can deduce entirely rational criteria for the solubility of certain embedding problems and determine the set of such solutions with restricted ramification. 3) Recent work on a correspondence between certain modular forms and certain two dimensional representations of Galois groups over $\mathbb{Q}$ uses a classification of linear representations via the corresponding projective representations and this of course involves central extensions. 4) The validity or failure of the Hasse norm principle is closely tied up with central extensions. 5) Finally the theory goes a good deal beyond what genus theory by itself can deliver in information about ideal class groups. We shall in particular be able to determine, again in terms of entirely rational criteria, all Abelian fields whose degree is a power of a prime ℓ and whose class number is not divisable by ℓ.

A group extension

$$S : 1 \to \Delta \to \Sigma \to \Gamma \to 1 \tag{3.1}$$

is central, if Δ is a subgroup of the centre $\mathrm{cent}(\Sigma)$ of Σ. A general group extension (3.1) has a maximal central quotient

$$1 \to \Delta/(\Delta,\Sigma) \to \Sigma/(\Delta,\Sigma) \to \Gamma \to 1 \tag{3.2}$$

Recall now the group theoretical discussion at the beginning of §2. The maximal quotient of (3.1) lifted from an Abelian extension is of course central, and the obstruction of (3.2) to be lifted from an Abelian extension is measured by the quotient group $\Delta_\cap(\Sigma,\Sigma)/(\Delta,\Sigma) = M_S$ associated with the sequence (3.1) Homomorphisms of sequences then give rise to homomorphisms of the groups M_S (i.e., we get a functor $S \longmapsto M_S$). A homomorphism $S \longrightarrow S'$ which is surjective on all terms and the identity on $\Gamma = \Gamma'$ induces a surjective homomorphism $M_S \to M_{S'}$. We quote a well known result.

<u>3.1 Proposition</u>. <u>There is an Abelian group</u> $M(\Gamma)$ <u>and for every exact sequence</u> S, <u>as in</u> (3.1), <u>there is a surjective homomorphism</u> $g_S : M(\Gamma) \to M_S$, <u>such that</u>

(i) <u>If</u> $S \to S'$ <u>is a homomorphism of extensions of</u> Γ <u>then the diagram</u>

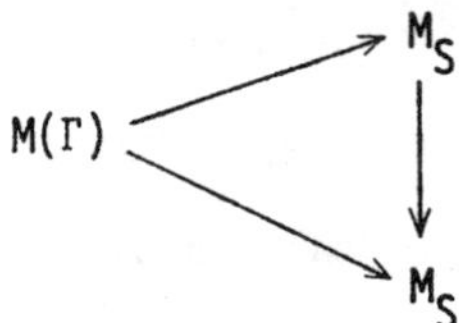

<u>commutes</u>.

(ii) <u>If</u> Σ <u>is a free group then</u> g_S <u>is an isomorphism</u>.

If Γ is finite then $M(\Gamma)$ is finite. In fact

$$M(\Gamma) \cong H_2(\Gamma, \mathbb{Z}) = \hat{H}^{-3}(\Gamma, \mathbb{Z}). \tag{3.3}$$

Thus $M(\Gamma)$ is the dual of the Schur multiplicator. There is also an analogous theory for profinite groups, in place of abstract groups.

Consider a tower of fields $N \supset L \supset K$, N/K and L/K Galois, with Galois groups $\mathrm{Gal}(N/L) = \Delta$, $\mathrm{Gal}(N/K) = \Sigma$, $\mathrm{Gal}(L/K) = \Gamma$. We are thus given a group extension (3.1). If $\Delta \subset \mathrm{cent}(\Sigma)$ we shall call N a central extension of L/K. In general, the fixed subfield N' of (Δ,Σ) will be the maximal central extension of L/K inside N. If moreover E is the maximal Abelian extension of K in N, we get, as a special case of Proposition 3.1, a surjective homomorphism - now denoted in terms of fields:

$$g_N : M(\Gamma) \to \Delta_\cap(\Sigma,\Sigma)/(\Delta,\Sigma) = \mathrm{Gal}(N'/EL). \tag{3.4}$$

Keeping this notation for an arbitrary tower $N \supset L \supset K$ as above, consider in particular an algebraic closure K^c of K; write $\Omega = \mathrm{Gal}(K^c/K)$, $\Lambda = \mathrm{Gal}(K^c/L)$ and put

$$\Lambda_\cap(\Omega,\Omega)/(\Lambda,\Omega) = \mathfrak{M}(L/K). \tag{3.5}$$

We then obtain a surjective homomorphism

$$g : M(\Gamma) \to \mathfrak{M}(L/K), \tag{3.6}$$

and for each N as above a surjective homomorphism

$$g'_N : \mathfrak{M}(L/K) \to \mathrm{Gal}(N'/EL), \tag{3.7}$$

so that the diagram

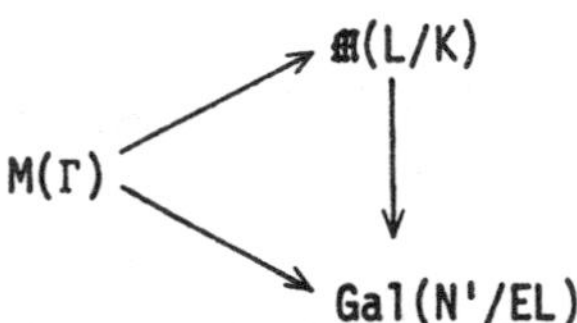

commutes. If g'_N is an isomorphism, we shall say that N <u>realizes</u> $\mathfrak{M}(L/K)$.

<u>3.2 Proposition</u>. <u>Suppose that</u> L/K <u>is of finite degree</u>. <u>Then there is an extension</u> N <u>of</u> L <u>of finite degree which realizes</u> $\mathfrak{M}(L/K)$.

<u>Proof</u>. $M(\Gamma)$ is finite (e.g., by (3.3)), hence so is $\mathfrak{M}(L/K)$. Thus, by (3.5), the closed subgroup (Λ,Ω) of $\Lambda_\cap(\Omega,\Omega)$ is also open in it. Hence there

is normal open subgroup Θ of Ω with $\Lambda_n(\Omega,\Omega)_n\Theta \subset (\Lambda,\Omega)$. The fixed field N of $\Lambda_n\Theta$ is of finite degree and will realize $\mathfrak{M}(L/K)$.

Remark. One can of course always impose the further condition on a field N, realizing $\mathfrak{M}(L/K)$, that it be a central extension of L/K.

From now on L/K is a Galois extension of finite degree of number fields or of local fields, with Galois group Γ. Recall the definition of $\hat{H}^{-1}(\Gamma,X)$ for a Γ-module X, written multiplicatively; it is the kernel of the "norm map" $x \longmapsto \Pi\, x^{\gamma}$ (γ running over Γ)modulo $X\,A(\Gamma)$, $A(\Gamma)$ the augmentation ideal of $\mathbb{Z}\Gamma$. We consider the reciprocity map over L, onto $\mathrm{Gal}(L^{ab}/L)$. By the commutativity of (1.9) it induces homomorphisms

$$\mathrm{Ker}\, N_{L/K} \to \mathrm{Gal}(L^{ab}/K^{ab}L).$$

Also it commutes with Galois action, hence induces homomorphisms

$$\left.\begin{array}{ll}\text{local} & L^{\times}A(\Gamma)\\ \text{global} & C(L)A(\Gamma)\end{array}\right\} \longrightarrow (\mathrm{Gal}(L^{ab}/L),\ \mathrm{Gal}(L^{ab}/K)).$$

Combining these, we end up with homomorphisms

$$\left\{\begin{array}{ll}\text{local:} & \hat{H}^{-1}(\Gamma,L^{\times})\\ \text{global:} & \hat{H}^{-1}(\Gamma,C(L))\end{array}\right\} \longrightarrow \mathfrak{M}(L/K). \tag{3.8}$$

3.3 Theorem. The map g of (3.6) and the maps (3.8) are all isomorphisms.

Proof. Let L' be the maximal central extension of L/K. Then

$$\mathfrak{M}(L/K) = \mathrm{Gal}(L'/K^{ab}L).$$

We first look at the local map (3.8). The inverse image of $\mathrm{Gal}(L^{ab}/K^{ab}L)$ under the reciprocity map is $\mathrm{Ker}\, N_{L/K}$. But the latter group is compact. As its image is dense, we get an isomorphism $\mathrm{Ker}\, N_{L/K} \cong \mathrm{Gal}(L^{ab}/K^{ab}L)$. The inverse image of $\mathrm{Gal}(L^{ab}/L')$ is $L^{\times}A(\Gamma)$. This shows that indeed the local map in

(3.8) is an isomorphism.

In the global case the reciprocity map is surjective, with kernel the connected component $D(L)$ of $C(L)$. We know (recall Lemma 2.2) that the inverse image of $\mathrm{Gal}(L^{ab}/K^{ab}L)$ is the subgroup $(\mathrm{Ker}\ N_{L/K})\cdot D(L)$ of $C(L)$, and this group maps surjectively onto $\mathrm{Gal}(L^{ab}/K^{ab}L)$. The inverse image of $\mathrm{Gal}(L^{ab}/L')$ is $(C(L)A(\Gamma))\cdot D(L)$. Thus we get an isomorphism

$$(\mathrm{Ker}\ N_{L/K})\cdot D(L)/(C(L)A(\Gamma))\cdot D(L) \cong \mathfrak{M}(L/K).$$

But clearly the group on the left is isomorphic to $\mathrm{Cok}[\hat{H}^{-1}(\Gamma,D(L)) \to \hat{H}^{-1}(\Gamma,C(L)]$. However, $\hat{H}^{-1}(\Gamma,D(L)) = 1$. Hence the map (3.8) is an isomorphism in the global case.

We know that the map g of (3.6) is surjective, and the groups involved are finite. We have to show that their orders are the same. In fact by the first part of the theorem, the groups $\hat{H}^{-1}(\Gamma,L^{\times})$ (local case) or $\hat{H}^{-1}(\Gamma,C(L))$ (global case) are isomorphic to $\mathfrak{M}(L/K)$. By class field theory they are also isomorphic to $\hat{H}^{-3}(\Gamma,\mathbb{Z})$, and this last group is isomorphic to $M(\Gamma)$ (see (3.3)).

Remark 1. Let L/K be a Galois extension of number fields with Galois group Γ. For every prime $\mathfrak{p}$ of K choose a prime $\mathfrak{P}$ of L above it and denote the Galois group of $L_{\mathfrak{P}}/K_{\mathfrak{p}}$ by $\Gamma_{\mathfrak{p}}$. This group is cyclic, and so $\hat{H}^{-1}(\Gamma_{\mathfrak{p}},L_{\mathfrak{P}}^{*}) = 1$ for all primes $\mathfrak{p}$ except possibly the finite, ramified ones. Denoting by Π^{*} the product over the latter, we have therefore an isomorphism

$$\Pi^{*}\ \hat{H}^{-1}(\Gamma_{\mathfrak{p}},L_{\mathfrak{P}}) \cong \hat{H}^{-1}(\Gamma,J(L)),$$

and we get a commutative diagram

$$\begin{array}{ccc} \hat{H}^{-1}(\Gamma,J(L)) & \to & \hat{H}^{-1}(\Gamma,C(L)) \\ \wr\Vert & & \wr\Vert \\ \Pi^{*}\ \mathfrak{M}(L_{\mathfrak{P}}/K_{\mathfrak{p}}) & \to & \mathfrak{M}(L/K) \end{array} \qquad (3.9)$$

where the top row comes from the map $J(L) \to C(L)$ and the bottom row from the maps $\Gamma_{\mathfrak{p}} \to \Gamma$.

Remark 2. Let N be a central extension of L/K (number field case). Then the diagram

$$\begin{array}{ccc} H^{-1}(\Gamma,C(L)) & = & \mathfrak{M}(L/K) \\ \downarrow & \swarrow & \\ \mathrm{Gal}(N/L) & & \end{array} \tag{3.10}$$

commutes where the column is given by the reciprocity map. Thus N realizes $\mathfrak{M}(L/K)$ precisely if this column is injective.

Duality. Let in the sequel L/K be a Galois extension of number fields with Galois group Γ. The dual norm $\check{N}_{L/K}$ maps $C(K)^{\dagger}$ into $(C(L)^{\dagger})^{\Gamma}$. Here the superscript Γ denotes fixed points. Clearly $\check{N}_{L/K}$ maps $C(K)^{\dagger}$ into $(C(L)/\mathrm{Ker}\, N_{L/K})^{\dagger}$, viewed as a subgroup of $C(L)^{\dagger}$. But if $\phi \in (C(L)/\mathrm{Ker}\, N_{L/K})^{\dagger}$, we may define ϕ' on $\mathrm{Im}\, N_{L/K}$ by $\phi'(N_{L/K}c) = \phi(c)$. Now $\phi \in \Phi(E/L)$ for some extension E of L, which we may take to be Galois over K. Thus $\phi'(\mathrm{Im}\, N_{E/K}) = 1$, and so we may extend ϕ' to a character ϕ'' in $\Phi(E/K)$. But then $\phi = \check{N}_{L/K}\phi''$. Thus indeed

$$\mathrm{Im}\, \check{N}_{L/K} = (C(L)/\mathrm{Ker}\, N_{L/K})^{\dagger}.$$

Obviously

$$(C(L)^{\dagger})^{\Gamma} = (C(L)/C(L)A(\Gamma))^{\dagger}.$$

We thus get an exact sequence

$$1 \to \Phi(L/K) \to C(K)^{\dagger} \xrightarrow{\check{N}_{L/K}} (C(L)^{\dagger})^{\Gamma} \to \Psi(L/K) \to 1, \tag{3.11}$$

where $\mathrm{Cok}\, \check{N}_{L/K}$ may be identified as a dual:

$$\Psi(L/K) = (\hat{H}^{-1}(\Gamma,C(L)))^{\dagger}. \tag{3.12}$$

If now N is a cental extension of L/K, and so $\Phi(N/L) \subset (C(L)^{\dagger})^{\Gamma}$ we obtain from (3.11) a map

$$\Phi(N/L) \to \Psi(L/K), \tag{3.13}$$

and we then get (see remark 2 above)

3.4 Proposition. *The central extension* N *of* L/K *will realize* $\mathfrak{M}(L/K)$ *if and only if the map in* (3.13) *is surjective*.

Next if $\mathfrak{p}$ is a prime divisor of K, and $\mathfrak{P}$ a prime divisor of L above $\mathfrak{p}$ we get, corresponding to (3.11), a local exact sequence

$$1 \to \Phi(L_{\mathfrak{P}}/K_{\mathfrak{p}}) \to (K_{\mathfrak{p}}^{\times})^{\dagger} \to ((L_{\mathfrak{P}}^{\times})^{\dagger})^{\Gamma_{\mathfrak{p}}} \to \Psi(L_{\mathfrak{P}}/K_{\mathfrak{p}}) \to 1 \tag{3.14}$$

where $\Gamma_{\mathfrak{p}} = \mathrm{Gal}(L_{\mathfrak{P}}/K_{\mathfrak{p}})$. But (see Remark 1 above) the group $\Psi(L_{\mathfrak{P}}/K_{\mathfrak{p}})$ is trivial for all non-ramified finite prime divisors $\mathfrak{p}$, and all infinite primes. Localization yields a homomorphism of the sequence (3.11) into the sequence (3.14). In particular we get a homomorphism

$$\Psi(L/K) \to \Pi^{*}\Psi(L_{\mathfrak{P}}/K_{\mathfrak{p}}), \tag{3.15}$$

where Π^{*} stands again for the product over those finite primes of K, which ramify in L.

Relation to the embedding problem. Let X be an Abelian group (with discrete topology and viewed as a trivial Γ-module). The central extensions

$$1 \to X \to Y \to \Gamma \to 1 \tag{3.16}$$

are then classified by $H^2(\Gamma,X)$. Associating with a given extension (3.16) the corresponding map $M(\Gamma) \longrightarrow X_n(Y,Y) \subset X$, we get a homomorphism

$$H^2(\Gamma,X) \to \mathrm{Hom}(M(\Gamma),X). \tag{3.17}$$

The spectral sequence associated with the quotient group Γ of $\mathrm{Gal}(L'/K)$ (where L' is the maximal central extension of L/K) gives rise to a homomorphism

$$\mathrm{Hom}(\mathrm{Gal}(L'/L),X) = H^1(\mathrm{Gal}(L'/L),X) \to H^2(\Gamma,X). \tag{3.18}$$

Explicitly one associates with a given homomorphism h the class of the bottom row in the commutative diagram with exact rows

$$\begin{array}{ccccccccc} 1 & \to & \mathrm{Gal}(L'/L) & \to & \mathrm{Gal}(L'/K) & \to & \Gamma & \to & 1 \\ & & \downarrow h & & \downarrow & & \downarrow & & \\ 1 & \to & X & \to & Y & \to & \Gamma & \to & 1 \end{array} \tag{3.19}$$

The embedding problem is the problem of finding the image of the map (3.18). Observe that, given h as in (3.19), the diagram

$$\begin{array}{ccc} M(\Gamma) \cong \mathfrak{m}(L/K) & \subset & \mathrm{Gal}(L'/L) \\ & \searrow & \downarrow h \\ & & X \end{array} \tag{3.20}$$

will commute.

We now specialize the above to $X = \mu$, the group of complex roots of unity with discrete topology (and trivial Γ-action).

<u>3.5 Theorem</u>. <u>For</u> $X = \mu$ <u>the map</u> (3.18) <u>is surjective</u>.

<u>Proof</u>. It is a well-known fact of group theory that, for $X = \mu$, the map (3.17) is an isomorphism. Composing (3.18) and (3.17) we get a commutative diagram

$$\begin{array}{ccc} \mathrm{Hom}(\mathrm{Gal}(L'/L),\mu) & \longrightarrow & \mathrm{Hom}(M(\Gamma),\mu) \\ \wr\| & & \wr\| \\ (C(L)^{\dagger})^{\Gamma} & \longrightarrow & \Psi(L/K) = \mathrm{Hom}(\mathfrak{m}(L/K),\mu). \end{array}$$

Here the left-hand column is the isomorphism, given by identifying continuous characters of $C(L)$ of finite order with characters of a Galois group; the right-hand column is the inverse dual of the isomorphism $M(\Gamma) = \mathfrak{m}(L/K)$. The commutativity follows from that of (3.20). The bottom row is surjective, as part of the exact sequence (3.11). Therefore the top row is surjective. It follows now that (3.18) is surjective.

We shall reformulate this result in terms of representation theory. Let

$$T : \mathrm{Gal}(K^c/K) \to GL_m(\mathbb{C})$$

be a continuous representation of the absolute Galois group of K; here K^c is the algebraic closure of K and $\mathbb{C}$ has discrete topology. Composing this with the projection $P : GL_m(\mathbb{C}) \to PGL_m(\mathbb{C})$ we get a commutative diagram

$$\begin{array}{ccc} \mathrm{Gal}(K^c/K) & \longrightarrow & GL_m(\mathbb{C}) \\ & \searrow^{PT} & \downarrow \\ & & PGL_m(\mathbb{C}) \end{array} \tag{3.21}$$

where PT is a continuous projective representation.

<u>Corollary to Theorem 3.5</u>. <u>Every continuous projective representation</u>

$$S : \mathrm{Gal}(K^c/K) \to PGL_m(\mathbb{C})$$

<u>lifts to a continuous representation</u> T, <u>i.e.</u>, <u>so that</u> $S = PT$.

Indeed let L be the fixed field of $\mathrm{Ker}\, S$, whence $\mathrm{Im}\, S \simeq \Gamma$. Identifying these two groups, and observing that $\mathbb{C}^* = \mathrm{cent}(GL_m(\mathbb{C}))$, it is clear that S will lift if there is a commutative diagram (3.19) with $X = \mathbb{C}^*$. The reduction from this to $X = \mu$ is standard cohomology theory.

<u>Non-ramified extensions</u>. Again L/K is a Galois extension of number fields, Γ the Galois group. We write $Z(L/K)$ and $Z_+(L/K)$ for the maximal central extension of L/K which is non-ramified, or non-ramified at all finite primes, respectively. $Z(L/K)$ is the <u>central class field</u>, $Z_+(L/K)$ the <u>narrow central class</u> field of L/K. We shall keep the discussion in terms of $Z(L/K)$, except where there is a special interest to look at both fields. The extension of results from $Z(L/K)$ to $Z_+(L/K)$ elsewhere is left to the reader. We can view $Z(L/K)$ as the maximal non-ramified extension of L inside the maximal central extension of L/K, or as the maximal central extension of L/K inside the absolute class field $L^{(1)}$ of L. From the latter point of view the reciprocity isomorphism $\mathcal{C}l(L) \cong \mathrm{Gal}(L^{(1)}/L)$ yields an isomorphism

$$\mathcal{C}l(L)/\mathcal{C}l(L)A(\Gamma) \cong \mathrm{Gal}(Z(L/K)/L). \tag{3.22}$$

Of course $G(L/K) \subset Z(L/K)$.

3.6 Proposition. The reciprocity map yields an isomorphism

$$\mathfrak{H}(L/K)/\mathfrak{Cl}(L)A(\Gamma) \cong \mathrm{Gal}(Z(L/K)/G(L/K))$$

$$= \mathrm{Im}[\mathfrak{M}(L/K) \to \mathrm{Gal}(Z(L/K)/L)]$$

Proof. The isomorphism is obtained by combining (3.22) with the reciprocity isomorphism $\mathfrak{H}(L/K) \cong \mathrm{Gal}(L^{(1)}/G(L/K))$. The identification of the Galois group with the image of $\mathfrak{M}(L/K)$ follows from the commutativity of (3.10). One can also prove this directly by showing that

$$\mathrm{Im}[\hat{H}^{-1}(\Gamma,C(L)) \to \hat{H}^{-1}(\Gamma,\mathfrak{Cl}(L))] = \mathfrak{H}(L/K)/\mathfrak{Cl}(L)A(\Gamma). \tag{3.23}$$

This is a consequence of the description of $\mathfrak{H}(L/K)$ in terms of ideles (see e.g., Lemma 2.3). On the other hand, with our proof of the Proposition, as above, (3.23) becomes a Corollary.

We now mention a number of Lemmas, which give the background for the applications of the theory of central class fields.

3.7 Lemma. Exponentiation by $[L : K]$ annihilates the group $\mathfrak{H}(L/K)/\mathfrak{Cl}(L)A(\Gamma)$, and if $\mathfrak{Cl}(K) = 1$, also the group $\mathfrak{Cl}(L)/\mathfrak{Cl}(L)A(\Gamma)$.

Proof. Both groups are subgroups of $\hat{H}^{-1}(\Gamma,\mathfrak{Cl}(L))$, see (3.23) for the first of these.

3.8 Lemma. Let L be the composite of Galois extensions L_1 and L_2 of K, of coprime degrees, with Galois groups Γ_1, and Γ_2, respectively. Then

$$\mathfrak{H}(L/K)/\mathfrak{Cl}(L)A(\Gamma) \cong ([\mathfrak{H}(L_1/K)/\mathfrak{Cl}(L_1)A(\Gamma_1)] \times [\mathfrak{H}(L_2/K)/\mathfrak{Cl}(L_2)A(\Gamma_2)]).$$

If $\mathfrak{Cl}(K) = 1$ also

$$\mathfrak{Cl}(L)/\mathfrak{Cl}(L)A(\Gamma) \cong ([\mathfrak{Cl}(L_1)/\mathfrak{Cl}(L_1)A(\Gamma_1)] \times [\mathfrak{Cl}(L_2)/\mathfrak{Cl}(L_2)A(\Gamma_2)]).$$

The proof is analogous to that of 2.8.

3.9 Lemma. Suppose L is its own genus field, or that $\mathfrak{Cl}(K) = 1$. Let

$[L : K]$ _be a power of a prime_ ℓ. _Then_

$$\mathfrak{Cl}(L)/\mathfrak{Cl}(L)A(\Gamma) = 1 \quad \text{if and only if} \quad \mathfrak{Cl}_\ell(L) = 1.$$

Proof. By Lemma 3.7, $\mathfrak{Cl}(L)/\mathfrak{Cl}(L)A(\Gamma)$ is an ℓ-group, so coincides with $\mathfrak{Cl}_\ell(L)/\mathfrak{Cl}_\ell(L)A(\Gamma)$. As the operating group Γ is an ℓ-group this quotient of $\mathfrak{Cl}_\ell(L)$ will vanish if and only if $\mathfrak{Cl}_\ell(L)$ does.

This lemma will lead to one of our main applications. The other one is based on the next lemma.

3.10 Lemma. _Let_ L/K _be cyclic and let_ M _be the genus field_ $G(L/K)$ _of_ L/K, _and_ N _the maximal Abelian extension of_ L _in_ $Z(M/K)$. _Then_

$$\mathfrak{H}(L/K)/\mathfrak{H}(L/K)^{1-\sigma} \cong \mathrm{Gal}(N/M)$$

where $\langle\sigma\rangle = \mathrm{Gal}(L/K)$.

Proof. Immediately from inspection of the following diagram:

Fields	Groups
$Z(M/K)$	$1 = (\Sigma,(\Sigma,\Delta))$
N	$(\Delta,\Delta) = (\Delta,\Delta)(\Sigma,(\Sigma,\Delta))$
M	$(\Sigma,\Sigma) = (\Sigma,\Delta)$
L	Δ
K	Σ

$\Sigma/\Delta = \langle\sigma\rangle$.

Relation to the Hasse norm principle. The group $K^\times \cap N_{L/K}(J(L))$ is the group of norm residues of $L^\times$ in $K^\times$. It contains the group $N_{L/K}(L^\times)$ of number norms. The Hasse norm principle states that the two groups coincide when L/K is cyclic. We consider the quotient group,

$$\mathfrak{N}(L/K) = K^\times \cap N_{L/K}(J(L))/N_{L/K}(L^\times) = \mathrm{Ker}[\hat{H}^\circ(\Gamma,L^\times) \to \hat{H}^\circ(\Gamma,J(L))], \tag{3.24}$$

(the number knot of Scholz) which measures the failure of the norm principle in the general case. Denote by $\mathfrak{U}(L/K)$, and $\mathfrak{U}_+(L/K)$ respectively the subgroup of $\mathfrak{N}(L/K)$, of classes mod $N_{L/K}(L^\times)$ which are represented by global units, and by totally positive global units, respectively.

<u>3.11 Theorem. There are exact sequences</u>

$$1 \to \mathfrak{U}_+(L/K) \to \mathfrak{N}(L/K) \to \mathfrak{H}_+(L/K)/\mathfrak{Cl}_+(L)A(\Gamma) \to 1,$$

$$1 \to \mathfrak{U}(L/K) \to \mathfrak{N}(L/K) \to \mathfrak{H}(L/K)/\mathfrak{Cl}(L)A(\Gamma) \to 1.$$

<u>Here the maps from</u> $\mathfrak{N}(L/K)$ <u>are as follows</u>: <u>Let</u> $a \in K^{\times}$, $a = N_{L/K}(\alpha)$, $\alpha \in J(L)$. <u>Then take the class of</u> (α) <u>in the appropriate group</u>.

<u>Proof</u>. (For the second sequence). We consider the commutative diagram with exact rows and columns, $Y(L)$ denoting the group of global units,

$$\begin{array}{ccccccccc}
 & & 1 & & 1 & & 1 & & \\
 & & \downarrow & & \downarrow & & \downarrow & & \\
1 & \longrightarrow & Y(L) & \longrightarrow & U(L) & \longrightarrow & U(L)/Y(L) & \longrightarrow & 1 \\
 & & \downarrow & & \downarrow & & \downarrow & & \\
1 & \longrightarrow & L^{\times} & \longrightarrow & J(L) & \longrightarrow & C(L) & \longrightarrow & 1 \\
 & & & & \downarrow & & \downarrow & & \\
 & & & & I(L) & \longrightarrow & \mathfrak{Cl}(L) & \longrightarrow & 1 \\
 & & & & \downarrow & & \downarrow & & \\
 & & & & 1 & & 1 & &
\end{array} \qquad (3.25)$$

We derive a commutative diagram with exact rows and columns (given by the cohomology sequences attached to (3.25))

$$\begin{array}{ccccccc}
\hat{H}^{-1}(\Gamma,U(L)) & \to & \hat{H}^{-1}(\Gamma,U(L)/Y(L)) & \to & \mathrm{Ker}[\hat{H}^{\circ}(\Gamma,Y(L)) \to \hat{H}^{\circ}(\Gamma,U(L))] & \to & 0 \\
\downarrow f & & \downarrow g & & \downarrow h & & \\
\hat{H}^{-1}(\Gamma,J(L)) & \to & \hat{H}^{-1}(\Gamma,C(L)) & \to & \mathfrak{N}(L/K) & \to & 0 \\
\downarrow & & & & & & \\
0 & & & & & &
\end{array}$$

The fact that f is surjective follows from Lemma 2.11, which told us that $\hat{H}^{-1}(\Gamma,I(L)) = 1$. Therefore $\mathrm{Cok}\ g \cong \mathrm{Cok}\ h$. But, via the right-hand column of (3.25), we see that $\mathrm{Cok}\ g = \mathrm{Im}[\hat{H}^{-1}(\Gamma,C(L)) \to H^{-1}(\Gamma,\mathfrak{Cl}(L))]$. Now apply (3.23). The verification that the map takes the form given is straight forward. It is also clear that $\operatorname{Im} h = \mathfrak{U}(L/K)$, as $\hat{H}^\circ(\Gamma, U(L)) \to \hat{H}^\circ(\Gamma, J(L))$ is injective.

Corollary. There is an exact sequence

$$1 \to \mathfrak{U}_+(L/K) \to \mathfrak{U}(L/K) \to \mathfrak{H}_+(L/K)/\mathfrak{Cl}_+(L)A(\Gamma) \to \mathfrak{H}(L/K)/\mathfrak{Cl}(L)A(\Gamma) \to 1.$$

Proof. The top sequence of 3.11 maps onto the bottom sequence, over the identity map of $\mathfrak{N}(L/K)$. Now apply the snake Lemma.

The rational case. We have seen in Theorem 3.5 that every central extension, of a Galois group $\mathrm{Gal}(L/K)$ of number fields, by the group $\mu \cong \mathbb{Q}/\mathbb{Z}$ is given by some character ϕ, i.e., continuous homomorphism $\mathrm{Gal}(L'/L) \to \mu$. Here we shall see that for base field $K = \mathbb{Q}$ the ramification at ϕ can be prescribed, to within given local conditions. Let $S(L)$ be the set of finite rational primes ramified in L. For each finite rational prime p, let $\mathfrak{P}$ be some prime divisor of L above p, and write $\Gamma_p = \mathrm{Gal}(L_{\mathfrak{P}}/\mathbb{Q}_p)$.

3.12 Theorem. Let $\psi \in \Psi(L/K)$. Suppose that, for each finite rational prime p, we are given a character $\phi(p) \in (U(L_{\mathfrak{P}})^{\dagger})^{\Gamma_p}$, so that (i) $\phi(p) = 1$ is the trivial character for almost all p (i.e., with at most finitely many exceptions,) and (ii) for all $p \in S(L)$, there is a character $\phi'(p) \in ((L_{\mathfrak{P}}^{\times})^{\dagger})^{\Gamma_p}$, whose restriction to $U(L_{\mathfrak{P}})$ is $\phi(p)$, and whose image in $\Psi(L_{\mathfrak{P}}/\mathbb{Q}_p)$ is the p-component ψ_p of ψ, under the localization map (3.15). Then there is a character $\phi \in (C(L)^{\dagger})^{\Gamma}$, whose image in $\Psi(L/\mathbb{Q})$ is the given ψ, and whose restriction to $U(L_{\mathfrak{P}})$ is $\phi(p)$, for all p.

ϕ is unique modulo $\check{\mathfrak{G}}_+(L/\mathbb{Q})$.

Remark. If we view characters as homomorphisms $\mathrm{Gal}(L'/L) \to \mu$ then the $\phi(p)$ are given as the restrictions of local characters to inertia groups.

Proof. If $p \notin S(L)$, i.e., $L_{\mathfrak{P}}/\mathbb{Q}_p$ is non-ramified then we can extend the given character $\phi(p)$ to a character $\phi'(p)$ in $((L_{\mathfrak{P}}^{\times})^{\dagger})^{\Gamma_p}$, and we assume that this has been done. Also in this case $\psi_p = 1$.

Now let $\phi'' \in (C(L)^{\dagger})^{\Gamma}$ have image ψ. Then for each finite p, $\phi'(p)\phi_{\mathfrak{P}}''^{-1}$ has trivial image in $\Psi(L_{\mathfrak{P}}/\mathbb{Q}_p)$, hence is of form $\check{N}_{L_{\mathfrak{P}}/\mathbb{Q}_p}\theta(p)$, $\theta(p) \in (\mathbb{Q}_p^{\times})^{\dagger}$.

Moreover we may assume that $\theta(p)$ is non-ramified whenever $\phi'(p)\phi''^{-1}_{\mathfrak{p}}$ is, i.e., at almost all p. In view of the isomorphism $C(\mathbb{Q})^{\dagger} \cong U_{fin}(\mathbb{Q})^{\dagger}$, there is a unique $\theta \in C(\mathbb{Q})^{\dagger}$, whose restriction to $U(\mathbb{Q}_p)$ coincides with that of $\theta(p)$, for all p. It follows that $(\check{N}_{L/\mathbb{Q}}\theta)_{\mathfrak{p}}$ differs from $\check{N}_{L_{\mathfrak{p}}/\mathbb{Q}_p}\theta(p) = \phi'(p)\phi''^{-1}_{\mathfrak{p}}$ by a non-ramified character. In other words, $\check{N}_{L/\mathbb{Q}}\theta\cdot\phi''$ coincides, for each p, with $\phi(p)$ on $U(L_{\mathfrak{p}})$. On the other hand, ψ is still the image of $\check{N}_{L/\mathbb{Q}}\theta\cdot\phi''$. This then is a character ϕ as required.

Given a further character ϕ_1, we see that ϕ and ϕ_1 will both satisfy the stated conditions if, and only if, $\phi_1\phi^{-1}$ is non-ramified at all finite prime divisors and moreover $\phi_1\phi^{-1}$ lies in

$$\mathrm{Ker}[(C(L)^{\dagger})^{\Gamma} \to \Psi(L/K)] = \mathrm{Im}\, \check{N}_{L/K}.$$

But this is the same as saying that $\phi_1\phi^{-1} \in \check{\mathfrak{G}}_{+}(L/K)$.

We have another related theorem. Let S be a finite set of finite rational primes containing $S(L)$. We define $C(\mathbb{Q})^{\dagger,S}$ to be the subgroup of $C(\mathbb{Q})^{\dagger}$ of characters which are non-ramified at all primes in S, and $C(L)^{\dagger}_{S}$ the subgroup of $C(L)^{\dagger}$ of characters which are non-ramified at all finite primes of L lying above those not in S. We have

<u>3.13 Theorem</u>. (i) <u>If</u> $\theta \in C(\mathbb{Q})^{\dagger,S}$ <u>and</u> $\check{N}_{L/\mathbb{Q}}\theta$ <u>is non-ramified then</u> $\theta = 1$. <u>In particular</u> $\check{N}_{L/\mathbb{Q}}$ <u>is injective on</u> $C(\mathbb{Q})^{\dagger,S}$.

(ii) $(C(L)^{\dagger})^{\Gamma} = (C(L)^{\dagger}_{S})^{\Gamma} \times (\check{N}_{L/\mathbb{Q}}(C(\mathbb{Q})^{\dagger,S}))$.

<u>Proof</u>. (i) In the usual notation, $N_{L_{\mathfrak{p}}/\mathbb{Q}_p}$ defines a surjection on local units, if p is a finite prime not in S. Thus if $\check{N}_{L/\mathbb{Q}}\theta$ is non-ramified, then θ is non-ramified outside S. But by hypothesis θ is non-ramified in S. Hence $\theta = 1$.

(ii) Let $\phi \in (C(L)^{\dagger})^{\Gamma}$. As in the proof of Theorem 3.12, we construct a character $\theta \in C(\mathbb{Q})^{\dagger,S}$ such that $\phi\cdot\check{N}_{L/\mathbb{Q}}\theta^{-1}$ is non-ramified outside S. This shows that the group on the left on (ii) is indeed generated by the two groups on the right. That their product is direct follows then from part (i).

Let $\phi \longmapsto \phi'$ denote the projection $(C(L)^{\dagger})^{\Gamma} \to (C(L)_S^{\dagger})^{\Gamma}$ given by the direct product in the above theorem. We have immediately

Corollary 1. *For any subgroup* Φ *of* $(C(L)^{\dagger})^{\Gamma}$ *the diagram*

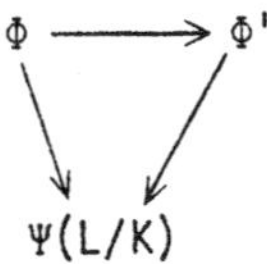

$$\begin{array}{ccc} \Phi & \longrightarrow & \Phi' \\ & \searrow \quad \swarrow & \\ & \Psi(L/K) & \end{array}$$

commutes.

Let in particular $\Phi = \Phi(M/L)$, where M is of finite degree over L, and M realizes $\mathfrak{M}(L/\mathbb{Q})$. Then $\Phi' = \Phi(M'/L)$, where M' is of finite degree over L, realizes $\mathfrak{M}(L/\mathbb{Q})$ and the only finite rational primes ramified in M' will lie in S. In other words, we get

Corollary 2. $\mathfrak{M}(L/\mathbb{Q})$ *is realized by a central extension* M *of* $L/\mathbb{Q}$, *of finite degree and ramified over* $\mathbb{Q}$ *only at the primes in* $S(L)$ (*and possibly at the infinite prime*).

Finally we come back to the embedding problem, using again the earlier notations, see (3.16)-(3.19). We now take

$$X = \mu_n = \frac{1}{n}\mathbb{Z}/\mathbb{Z} \subset \mathbb{Q}/\mathbb{Z} = \mu.$$

The group $\mathrm{Hom}(\mathrm{Gal}(L'/L),\mu_n)$ - where again L' is the maximal central extension of $L/\mathbb{Q}$ - may be identified with the subgroup of $(C(L)^{\dagger})^{\Gamma}$ of characters of order dividing n.

Corollary 3. *Let* $\phi \in \mathrm{Hom}(\mathrm{Gal}(L'/L),\mu_n)$. *The image of* ϕ *in* $H^2(\Gamma,\mu_n)$ *is the same as the image of* ϕ'. *Thus every realizable class* $c \in H^2(\Gamma,\mu_n)$ (*i.e.*, *every class in* $\mathrm{Im}[\mathrm{Hom}(\mathrm{Gal}(L'/L),\mu_n) \to H^2(\Gamma,\mu_n)]$) *is image of a character which is non-ramified outside* S.

Proof. We have to show that if $\theta \in C(\mathbb{Q})^{\dagger,S}$, and $\check{N}_{L/\mathbb{Q}}\theta$ is of order dividing n then it lies in the kernel of the map into $H^2(\Gamma,\mu_n)$. By Theorem

3.13, θ itself is of order dividing n. In terms of Galois groups, $\check{N}_{L/\mathbb{Q}}$ corresponds to restriction of characters from $\mathrm{Gal}(L'/\mathbb{Q})$ to $\mathrm{Gal}(L'/L)$. Thus in the diagram (3.19) (with $X = \mu_n$) the map h is now the restriction to $\mathrm{Gal}(L'/L)$ of the map $\theta : \mathrm{Gal}(L'/K) \to \mu_n$. This implies that the bottom sequence splits - this is what we had to show.

4. Maximal Quasicentral Extensions, Maximal ℓ-Extensions and Maximal Class Two Extensions

Our principal aim here is the description of certain Galois groups over $\mathbb{Q}$. We shall however also derive analogous results for local fields, and we begin with the required group theoretic background.

The theory of the functor $M(\Gamma)$ (see Proposition 3.1) can be developed in the context of profinite groups, rather than of abstract, i.e., of discrete groups. The restatement of results is obvious, and in any case for finite groups the two points of view coincide. Thus if $\Gamma \cong \Omega/\Lambda$, and Ω is a free profinite group, then

$$M(\Gamma) \cong \Lambda_{\cap}(\Omega,\Omega)/(\Lambda,\Omega).$$

Here of course commutator subgroups, such as (Λ,Ω), are to be understood topologically, i.e., (Λ,Ω) is the closure of the subgroup generated by the commutators.

In the context of a pair Ω, Σ, where Σ is a closed normal subgroup of a profinite group Ω we shall always write

$$\Sigma^{(1)} = (\Sigma,\Omega), \quad \Sigma^{(j+1)} = (\Sigma^{(j)},\Omega), \quad \Sigma^{(\infty)} = \bigcap_j \Sigma^{(j)}. \tag{4.1}$$

If $\Sigma^{(1)} = 1$ then of course Σ is a central subgroup of Ω, Ω is a central extension of Ω/Σ. If $\Sigma^{(\infty)} = 1$ then Σ is a _hyper central_ subgroup of Ω, Ω is a hyper central extension of Ω/Σ.

A profinite group Ω is said to be _multiplicator free_ if $M(\Omega) = 1$.

4.1 Proposition. Ω is multiplicator free, if and only if, for all open normal subgroups Σ of Ω, the map

$$M(\Omega/\Sigma) \to \Sigma_\cap(\Omega,\Omega)/(\Sigma,\Omega)$$

is an isomorphism. If Ω *is multiplicator free, then a surjection* $\Omega' \to \Omega$ *with central, or more generally hyper central kernel, which gives rise to an isomorphism* $\Omega'^{ab} \to \Omega^{ab}$*, is itself an isomorphism.*

Proof. Standard.

Let T be a nonempty set of primes. A T-primary profinite group is the inverse limit of T-primary finite groups. A finite group is T-primary if the primes dividing its order lie in T.

4.2 *Proposition. Suppose* Ω *is multiplicator free and let* Σ *be an open normal subgroup of* Ω.

(i) $\Omega/\Sigma^{(\infty)}$ *is multiplicator free, with hyper central subgroup* $\Sigma/\Sigma^{(\infty)}$.

(ii) *Let* T *contain all primes dividing the order of* Ω/Σ. *Let* $\Sigma/\Sigma_T^{(\infty)}$ *be the maximal* T*-primary quotient of* $\Sigma/\Sigma^{(\infty)}$. *Then* $\Sigma_T^{(\infty)}$ *is closed and normal in* Ω, *and* $\Omega/\Sigma_T^{(\infty)}$ *is multiplicator free, with hyper central subgroup* $\Sigma/\Sigma_T^{(\infty)}$.

Proof. The only assertions which are not immediately obvious are those referring to the multiplicator free property.

For (i) let Δ be open normal in $\Omega, \Delta \supset \Sigma^{(\infty)}$. Then $\Delta \supset \Sigma^{(j)}$, for some j, and hence $(\Delta,\Omega) \supset \Sigma^{(j+1)}$. This implies that the map

$$\Delta_\cap(\Omega,\Omega)/(\Delta,\Omega) \to \overline{\Delta}_\cap(\overline{\Omega},\overline{\Omega})/(\overline{\Delta},\overline{\Omega})$$

with $\overline{\Omega} = \Omega/\Sigma^{(\infty)}$, $\overline{\Delta} = \Delta/\Sigma^{(\infty)}$, is an isomorphism. Using Proposition 4.1 twice, we conclude that $\overline{\Omega}$ is multiplicator free.

For (ii) we may suppose that $\Sigma^{(\infty)} = 1$. Then Σ is pro-nilpotent, hence

$$\Sigma \cong (\Sigma/\Sigma_T^{\infty}) \times \Sigma_T^{(\infty)}$$

where $\Sigma_T^{(\infty)}$ is T'-primary, T' the complement of T. If now Δ is an open subgroup of $\Omega, \Delta \supset \Sigma_T^{(\infty)}$, then indeed Δ is T-primary, therefore $M(\Omega/\Delta) = \Delta_\cap(\Omega,\Omega)/(\Delta,\Omega)$ is T-primary, and so the surjection $\Omega \to \Omega^* = \Omega/\Sigma_T^{(\infty)}$ gives rise to an isomorphism

$$\Delta_\cap(\Omega,\Omega)/(\Delta,\Omega) \to \Delta_\cap^*(\Omega^*,\Omega^*)/(\Delta^*,\Omega^*)$$

where $\Delta^* = \Delta/\Sigma_T^{(\infty)}$. Using Proposition 4.1, the conclusion follows.

Let ℓ be a prime. A pro ℓ-group is the inverse limit of finite ℓ-groups. Such a group Λ is pro nilpotent, i.e., taking in (4.1) $\Omega = \Lambda$, $\Sigma = \Lambda$, we have $\Lambda^{(\infty)} = 1$. The maximal pro ℓ-quotient of a profinite group Ω is $\varprojlim \Omega/\Sigma$, where Σ runs through the open normal subgroups of Σ of ℓ-power index. From Proposition 4.2 we get the

Corollary. *The maximal pro ℓ-quotient of a multiplicator free profinite group Ω is multiplicator free.*

Recall that the notion of a generating set of a profinite group Ω is always to be interpreted topologically. We restrict ourselves to finite such sets. If in particular Ω is a pro ℓ-group, then any generating set S of $\Omega^{ab} = \Omega/\Omega^{(1)}$ lifts to a generating set of Ω. Let S be such. A non-trivial relation in Ω^{ab}, expressed in terms of S, is one which does not follow from the fact that Ω^{ab} is Abelian. Every such relation lifts to a relation holding in Ω. Let now $R^{(1)}$ be a set of non-trivial defining relations of Ω^{ab} as an Abelian group. Lift each relation in $R^{(1)}$ to a relation in Ω, given in terms of S, the latter viewed now as a generating set of Ω. Let R be the resulting set.

4.3 Proposition. *Let Ω be a multiplicator free pro ℓ-group, with $R^{(1)}$ and R as above. Then R is a set of defining relations of Ω as a pro ℓ-group.*

(In other words, Ω is the quotient of the free pro ℓ-group on S modulo the relations given by R.)

Proof. Let Ω' be the pro ℓ-group on the generating set S, with defining relations R. The identity map on S extends to a continuous surjection $\Omega' \to \Omega$, which induces an isomorphism $\Omega'^{ab} \simeq \Omega^{ab}$. The kernel is hyper central as Ω' is a pro ℓ-group, i.e., is pro nilpotent. By Proposition 4.1, $\Omega' \to \Omega$ is an isomorphism.

If, instead of lifting $R^{(1)}$ to R, we merely lift it to a set $R^{(k)}$ of relations of $\Omega/\Omega^{(k)}$, the same conclusion clearly holds, i.e., we have the

Corollary. Let Ω be a multiplicator free pro ℓ-group, with $R^{(1)}$ and $R^{(k)}$ as above. Then $R^{(k)}$ is a set of defining relations of $\Omega/\Omega^{(k)}$ as a pro ℓ-group of nilpotency class k.

(Recall that a group Λ is said to be nilpotent of class k if $\Lambda^{(k)} = 1$.)

Local fields. Here F is a local field of residue class characteristic p, of finite degree over $\mathbb{Q}_p$. By Theorem 3.3 we know that the absolute Galois group Ω_F of F, i.e., the group of the algebraic closure of F, over F, is multiplicator free. Applying the preceding propositions we obtain other multiplicator free Galois groups. We shall concentrate on the Galois group

$$\Omega_F\{\ell\} = \mathrm{Gal}(F^{(\ell)}/F), \tag{4.2}$$

where ℓ is a prime, $F^{(\ell)}$ is the maximal ℓ-extension of F, i.e., the composite of all Galois extensions of ℓ power degree of F (inside an algebraic closure) and so $\Omega_F\{\ell\}$ is the maximal pro ℓ-quotient of Ω_F.

Let in the sequel

$$m = \ell^r \tag{4.3}$$

be the order of the group of ℓ-power roots of unity in F. The pro ℓ-completion $P_\ell(F)$ of F^* is the product of s copies of $\mathbb{Z}_\ell$, where

$$s = \begin{cases} 1, & \text{if } \ell \neq p, \\ 1 + [F : \mathbb{Q}_p], & \text{if } \ell = p, \end{cases} \tag{4.4}$$

and further, in the case $m > 1$, a cyclic group of order m generated by a primitive m-th root y of 1. By local class field theory, this structure transfers itself isomorphically to the group

$$\Omega_F\{\ell\}^{ab} = \mathrm{Gal}(E/F), \tag{4.5}$$

where here E is the maximal Abelian ℓ-extension of F. By the Corollary to Proposition 4.2, and Proposition 4.3 we now have

4.4 Theorem. (i) (Theorem of Safarevic). If $m = 1$, then $\Omega_F\{\ell\}$ is the free pro ℓ-group on s generators.

(ii) If $m > 1$, then $\Omega_F\{\ell\}$ has a minimal generating set σ_j $(j = 0,1,\ldots,s)$, where $\sigma_0^{ab} = (E/F,y)$, with one defining relation, which is lifted from

$$(\sigma_0^{ab})^m = 1.$$

Here, and in the sequel, the image of an element γ of some group Γ in Γ^{ab} is to be denoted by γ^{ab}.

The exact form of the relation in (ii) can be given. We shall content ourselves here with doing this mod $\Omega_F\{\ell\}^{(2)}$, i.e., in class two. For this we shall need the local m-th power residue symbol. To define this all we need to assume for the moment is that F^* contains a torsion subgroup μ_m of order m.

Let L be the maximal Abelian extension of F, whose Galois group is of exponent dividing m. Then the m-th radical $\alpha = a^{1/m}$ of any element a of $F^\times$ lies in L, and L is generated by these. Moreover the element

$$\alpha^{(L/K,b)-1} = (a,b), \quad b \in F^\times$$

lies in μ_m and only depends on a and b, not on the choice of α. In fact the m-th power residue symbol (a,b) only depends on a and $b \bmod F^{\times m}$. It has the following properties

(i) $a,b \longrightarrow (a,b)$ defines a bimultiplicative non degenerate pairing $(F^\times/F^{\times m}) \times (F^\times/F^{\times m}) \to \mu_m$.

(ii) $(a,b) = 1$ if, and only if, $b \in N_{F(\alpha)/F}(F(\alpha)^\times)$.

(iii) $(a,-a) = 1 = (a,b)(b,a)$.

(iv) If F' is an extension field of F of finite degree, $a \in F^\times$, $b \in F'^\times$, then

$$(a,b)_{F'} = (a,N_{F'/F}b)_F.$$

From now on we shall again assume that $m > 1$ is the highest power of ℓ so that $\mu_m \subset F^\times$, and y will always be fixed as a generator of μ_m. In considering the symbol $(\ , \ )$ we may use the same letters for elements of $F^\times$ and for their classes in $F^\times/F^{\times m}$. Note that the latter group is the direct product of $s + 1$ cyclic groups of order m, and a set of generators of these - or a set of elements of $F^\times$ representing these generators - will be called a basis of $F^\times/F^{\times m}$.

Now consider first the case when $\ell = p$, and so $s = [F : \mathbb{Q}_p] + 1$. We define the notion of a quasi-symplectic basis of $F^\times/F^{\times m}$, as follows: We always assume that $y = a_0$.

(a) $p \neq 2$. Then s is odd. A basis $\{a_j\}$ $(j = 0,1,\ldots,s)$ of $F^\times/F^{\times m}$ is quasi-symplectic if for $i = 0,\ldots,(s-1)/2$,

$$(a_{2i},a_{2i+1}) = (a_{2i+1},a_{2i})^{-1} = y,$$
$$(a_j,a_k) = 1 \quad \text{otherwise.}$$

(b) $p = 2$, s odd (this includes the case when $4|m$). A basis $\{a_j\}$ $(j = 0,1,\ldots,s)$ of $F^\times/F^{\times m}$ is quasi-symplectic, if for $i = 0,\ldots,(s-1)/2$

$$(a_{2i},a_{2i+1}) = (a_{2i+1},a_{2i})^{-1} = y,$$
$$(a_1,a_1) = -1,$$
$$(a_j,a_k) = 1, \quad \text{otherwise.}$$

(c) $p = 2$, s even, hence $m = 2$, i.e., $y = -1$. A basis $\{a_j\}$, $(j = 0,1,\ldots,s)$ of $F^\times/F^{\times m}$ is quasi-symplectic if

$$(a_0,a_0) = -1$$
$$(a_{2i},a_{2i-1}) = (a_{2i-1},a_{2i}) = -1 \quad \text{for} \quad i = 1,\ldots,\tfrac{s}{2},$$
$$(a_j,a_k) = 1, \text{ otherwise.}$$

Then we have

4.5 Lemma. Quasi-symplectic bases exist.

Proof. We shall give the proof only in the case $p = 2$. The proof for odd p is similar, but easier.

First if $F = \mathbb{Q}_2$, then $\{-1,2,5\}$ is a quasi-symplectic basis. In fact

$$(-1,-1) = (2,5) = (5,2) = -1, \text{ all other values are } = 1. \tag{4.6}$$

More generally when s is even, i.e., $[F : \mathbb{Q}_2]$ is odd, then by property (iv) of the symbol $(\ ,\ )$ (with F, $\mathbb{Q}_2$ in place of F', F) again $(-1,-1)_F = -1$. The orthogonal complement H of $-1 \bmod F^{\times 2}$ consists of the classes of elements $a \bmod F^{\times 2}$ with $(-1,a) = 1$. As $(-1,-1) = -1$, the pairing $(\ ,\ )$ restricted to $H \times H$ is non degenerate, and by property (iii), $(a,a) = 1$ for all $a \in H$. Thus H is a symplectic space, i.e., has a symplectic basis $\{a_j\}$ $(j = 1,\ldots,s)$ as indicated. This deals with case (c).

Next if we still have $m = 2$, but s is odd, then the same argument as above gives $(-1,-1)_F = 1$. On the other hand, if $4|m$, so s is odd, then $(y,y) = (y,-1) = (y,y)^{m/2}$. Thus $(y,y^{1+m/2}) = 1$. But $1 + m/2$ is odd. So $(y,y) = 1$. This then holds whenever s is odd. Now choose a_1 with $(y,a_1) = y$, whence $(-1,a_1) = -1$. Therefore $(a_1,a_1) = -1$. The orthogonal complement H of $\{y,a_1\} \bmod F^{\times m}$ then admits $(\ ,\ )$ as a non-degenerate, alternating form, i.e., H is a symplectic space, and we may choose a symplectic basis $\{a_j\}$ $(j = 2,\ldots,s)$ as indicated.

Let now

$$\Lambda = \Omega/\Omega^{(2)} \quad \text{for} \quad \Omega = \Omega_F\{\ell\}. \tag{4.7}$$

Thus Λ is the Galois group of the maximal ℓ-extension of F of class two. We shall determine Λ by generators and relations, with a partial description of the generators in terms of $F^\times/F^{\times m}$. As before we let L be the maximal Abelian extension of F of exponent m.

<u>4.6 Theorem</u>. (Case $p = \ell$). <u>Let</u> $\{a_j\}$ <u>be a quasi symplectic basis of</u> $F^\times/F^{\times m}$. <u>Then</u> Λ <u>has a minimal generating set</u> $\{\sigma_j\}$ <u>in biunique correspondence with</u> $\{a_j\}$, <u>so that, for each</u> j, $(L/F,a_j)$ <u>is the image of</u> σ_j <u>in</u> $\mathrm{Gal}(L/F)$, <u>and so that</u> Λ <u>is the pro</u> ℓ-<u>group of class two with the one defining relation</u>

$$\sigma_0^m = \prod_i (\sigma_{2i},\sigma_{2i+1}), \quad (i = 0,\dots,(s-1)/2),\ s\ \underline{\text{odd}},$$

$$\sigma_0^m = \prod_i (\sigma_{2i-1},\sigma_{2i}), \quad (i = 1,\dots,s/2),\ s\ \underline{\text{even}}.$$

Proof. By the Corollary to Proposition 4.3 and by Theorem 4.4, the relation expressing σ_0^m in terms of commutators will be a defining relation. For simplicity, we stick to the case s odd. The proof works, with the appropriate change in notation, just as well for s even.

Any set $\{\sigma_j\}$ of elements, so that the image set $\{\sigma_j^{ab}\}$ in Λ^{ab} is a minimal generating set of Λ^{ab}, will be a minimal generating set of Λ. We can moreover postulate that the image of σ_j in $\mathrm{Gal}(L/F)$ should be $(L/F,a_j)$. Suppose then that we have such a set $\{\sigma_{j,k}\}$ (k a fixed second subscript) so that

$$\sigma_{0,k}^m \equiv \prod_i (\sigma_{2i,k},\sigma_{2i+1,k}) (\mathrm{mod}\ \Lambda^{(1)mp^k}). \tag{4.8}$$

By setting, for $j > 0$, and with suitable $t_{j,i,k}$,

$$\sigma_{j,k+1} = \sigma_{j,k} \prod_i \sigma_i^{t_{j,i,k}mp^k}, \quad \text{and} \quad \sigma_{0,k+1} = \sigma_{0,k}$$

(the product $\prod_i$ being taken in a given order) we get a congruence analogous to (4.8) but modulo $\Lambda^{(1)mp^{k+1}}$. In the limit we then obtain the required equation. We thus have to start with $k = 0$. Observe then that

$$\Lambda^{(1)m} = (\Delta,\Lambda),\ L \text{ the fixed field of } \Delta \subset \Lambda.$$

So we now work in $\Lambda/(\Delta,\Lambda) = \Lambda'$. We let $\{\gamma_j\}$ be elements of Λ', forming a minimal generating set of $\Lambda'^{ab} = \Lambda^{ab}$, and so that the image of γ_j in $\Lambda'/\Delta' = \mathrm{Gal}(L/F)$ is $(L/F,a_j)$ (where $\Delta' = \Delta/(\Delta,\Lambda)$). We shall then show that

$$\gamma_0^m = \prod_i (\gamma_{2i},\gamma_{2i+1}) \tag{4.9}$$

with the range of i as in the theorem. We know of course that we have some equation

$$\gamma_0^m = \prod_{j<k} (\gamma_j,\gamma_k)^{x_{jk}}, \tag{4.10}$$

and we have to determine the x_{jk} (in $\mathbb{Z}/m\mathbb{Z}$). Let now for every subscript j, Γ_j be the subgroup of Λ' generated by Δ' and the γ_k, for all $k \neq j$. Modulo (Γ_j,Γ_j) the commutator group (Λ',Λ') is the direct product of cyclic groups of order m with generators the cosets of (γ_j,γ_k), for all $k \neq j$. Thus the x_{jk} in (4.10), for the given j, and variable k, can already be determined by working modulo (Γ_j,Γ_j). In other words we have to establish a congruence

$$\gamma_0^m \equiv \prod_i (\gamma_{2i},\gamma_{2i+1}) \ (\mathrm{mod}(\Gamma_j,\Gamma_j)), \tag{4.11}$$

for each j.

Let L_j be the fixed field of Γ_j, N_j that of (Γ_j,Γ_j). Thus N_j is a central ℓ-extension of L_j over F, Abelian over L_j and

$$\begin{cases} L_j = F(\alpha_r),\ \alpha_r^m = a_r,\ a_r \text{ that element of the} \\ \text{given basis, with } (a_r,a_k) = 1 \text{ all } k \neq j. \end{cases} \tag{4.12}$$

For instance, if $j = 2i$, $i \geq 1$, then $r = 2i + 1$. To reformulate (4.11) in terms of $\mathrm{Gal}(N_j/F)$, we just have to replace each γ_v by an element $(L_j/F,a_v)$ lifted to $\mathrm{Gal}(N_j/F)$. The required equation then follows from the next proposition, together with computations of transfers of elements.

For the following proposition all we need is that m is some natural number for which F contains the primitive m-th roots of unity, whence the symbol $(a,b)_m = (a,b) \in \mu_m$ is defined. For simplicity sake let $a \in F$ be an element, so that $X^m - a$ is irreducible over F. Let α be a root. Write $e = 1$ if m is odd, $e = -1$ if m is even. Let M be an Abelian extension field of $F(\alpha)$, Galois over F, with Galois group $\mathrm{Gal}(M/F)$ nilpotent of class two. If $b \in F$ and M_1 is the maximal Abelian extension of F in M, lift the element $(M_1/F,b)$ to an element $(M/F,b) \in \mathrm{Gal}(M/F)$. Write v for the transfer (Verlagerung) $\mathrm{Gal}(F(\alpha)/F) \to \mathrm{Gal}(M/F(\alpha))$. Then we have

4.7 Proposition. Under the stated hypothesis we get the equations

$$((M/F,ae),\ (M/F,b)) = v((F(\alpha)/F,(a,b))),\ \text{all}\ \ b \in F,$$

connecting the commutators of the lifted Artin symbols $(M/F,\ \ )$ and the Artin symbols $(F(\alpha)/F,(a,b))$.

Proof. We shall evaluate $(M/F(\alpha),\alpha^{(M/F,b)-1})$ in two different ways. Firstly the Artin map preserves the action of a Galois group on the base field $F(\alpha)$. Therefore the above symbol can be written as $(M/F,b)^{-1}(M/F(\alpha),\alpha)(M/F,b)\cdot (M/F(\alpha),\alpha)^{-1} = ((M/F(\alpha),\alpha),\ (M/F,b))$. Moreover (cf. (1.9)) $(M/F(\alpha),\alpha) = (M/F,ae)$, as $ae = N_{F(\alpha)/F}(\alpha)$. On the other hand, of course $\alpha^{(M/F,b)-1} = (a,b)$. As the reciprocity map converts embeddings of fields into the transfer, we now get

$$(M/F(\alpha),(a,b)) = v((M/F,(a,b)).$$

This gives the required equation.

We now come to the easier case $p \neq \ell$. We let again $m > 1$ be the order of the ℓ-primary part of the torsion subgroup of F. If $N\mathfrak{p}_F$ is the absolute norm of the maximal ideal $\mathfrak{p}_F$, we have $N\mathfrak{p}_F \equiv 1 (\mathrm{mod}\ m)$. Let w be a primitive $N\mathfrak{p}_F - 1$ st root of unity, and let $w_1^m = w$, so $w_1 \in E$, the field defined in (4.5). Then $\mathrm{Gal}(E/F)$ is the direct product of a copy of $\mathbb{Z}_\ell$ generated by an element which acts as a Frobenius on $F(w_1)$, and of a cyclic group of order m generated by $(E/F,w)$. For the group Λ (cf. (4.7)) we now get

4.8 Theorem. Let σ be an element of Λ, so that

$$\sigma^{ab} = (E/F,w).$$

Let ω be an element of Λ which acts as a Frobenius element on $F(w_1)$. Then $\{\sigma,\omega\}$ is a minimal generating set of Λ, and Λ as a pro ℓ-group of class two is defined by the relation

$$\sigma^{N\mathfrak{p}_F-1} = (\sigma,\omega).$$

Proof. As $F(w_1)/F$ is non ramified, the norm map is surjective on the residue class fields. It follows that $w = N_{F(w_1)/F}(w_2)$, w_2 a root of unity. Therefore

$$\sigma^{ab} = (E/F(w_1), w_2).$$

As $w_2^{\omega} = w_2^{N\mathfrak{p}_F}$, we get from this, on lifting to Λ, the equation

$$\sigma^{N\mathfrak{p}_F - 1} = (E/F(w_1), w_2^{\omega-1}) = \sigma^{-1}\omega^{-1}\sigma\omega = (\sigma,\omega),$$

as required. By the Corollary to Proposition 4.3 this is then the defining relation of Λ as a pro ℓ-group of class two.

Theory over $\mathbb{Q}$. Let S be a set of finite rational primes. From Theorem 3.3 and the Corollary 2 to Theorem 3.13 we get

4.9 Theorem. The Galois group of the maximal extension of $\mathbb{Q}$, which is non ramified at all finite primes outside S is multiplicator free.

In the sequel we take S to be a finite set and ℓ again a fixed prime number. We shall now write $\Omega[S]$, or just Ω for the Galois group over $\mathbb{Q}$, of the maximal ℓ-extension of $\mathbb{Q}$, non ramified at all finite primes outside S. Then the maximal Abelian ℓ-extension $E = E[S]$ of $\mathbb{Q}$, non-ramified of all finite primes outside S has Galois group

$$\Omega^{ab} = \mathrm{Gal}(E/\mathbb{Q}) \quad \text{(new notation!)}. \tag{4.13}$$

Associated with certain primes p in S we introduce local elements w_p, which will also be viewed as rational ideles with components 1 outside p.

If $p \in S$, $p \equiv 1 (\mathrm{mod}\ \ell)$ then w_p is a primitive $p - 1$ st root of unity in $\mathbb{Q}_p$,

If $p = \ell \in S$, $\ell \neq 2$, then w_p is a generator (over $\mathbb{Z}_p$) of the group of units $\equiv 1 (\mathrm{mod}\ p)$ in $\mathbb{Z}_p$, e.g., $w_p = 1 + p$, (4.14)

If $p = \ell = 2 \in S$, then w_2 is a generator (over $\mathbb{Z}_2$) of the groups of units $\equiv 1 (\mathrm{mod}\ 4)$ in $\mathbb{Z}_2$, e.g., $w_2 = 5$.

These elements are to be kept fixed and will be referred to regularly in the sequel.

4.10 Theorem. *The group $\Omega = \Omega[S]$ has a minimal generating set $\{\tau_x\}$, where x runs through the primes in S of form $p \equiv 1 (\mathrm{mod}\ \ell)$, or $p = \ell$, and in the case $2 \in S$, an additional value $x = -1$, so that their images τ^{ab} in Ω^{ab} satisfy*

$$\tau_p^{ab} = (E/\mathbb{Q}, w_p)$$

and

$$\tau_{-1}^{ab} = (E/\mathbb{Q}, (-1)_2) \quad \text{if} \quad 2 \in S.$$

(Here $(-1)_2$ is the idele with component -1 at 2, and component 1 elsewhere).

A complete set of defining relations of Ω as a pro ℓ-group is lifted from the set of relations

$$(\tau_p^{ab})^{p-1} = 1 \quad \text{if} \quad p \in S,\ p \equiv 1(\mathrm{mod}\ \ell)$$

$$(\tau_{-1}^{ab})^2 = 1 \quad \text{if} \quad 2 \in S.$$

Proof. This follows from Theorem 4.9, Proposition 4.3 and the determination of $\Omega^{ab} = \mathrm{Gal}(E/\mathbb{Q})$, which is isomorphic to the pro ℓ-quotient of $\Pi_{p\in S} U(\mathbb{Q}_p)$, the latter being generated by the elements w_x as above. Note that the primes $p \in S$, $p \neq \ell$, $p \not\equiv 1(\mathrm{mod}\ \ell)$ make no contribution. From now on we shall assume that only primes $p \equiv 1(\mathrm{mod}\ \ell)$ or $p = \ell$ may be included in S.

The elements τ_x of Ω, and similarly their images in any quotient of Ω will be called *inertia generators*. They do indeed generate the inertia groups modulo commutators.

We shall describe the defining relations modulo $\Omega^{(2)}$, i.e., we shall completely determine the group

$$\Lambda[S] = \Lambda = \Omega/\Omega^{(2)} \quad (\Omega = \Omega[S]),$$

which is the Galois group of the maximal ℓ-extension of class two of $\mathbb{Q}$ non-ramified at all finite primes outside S. This determination will be arithmetically interesting, in that it will contain a description of the inertia groups. For each $p \in S$ we shall denote by $\Lambda_{(p),0}$ the inertia group in Λ of some prime divisor above p; this is unique modulo conjugacy. Hence both $\Lambda_{(p),0} \cap \Lambda^{(1)}$ and the image of $\Lambda_{(p),0} \bmod \Lambda^{(1)}$ are actually unique.

It will be convenient to denote by S' the set of subscripts associated with the given set S. Thus

$$\begin{cases} S' = S \cup \{-1\}, & \text{if} \quad 2 \in S, \\ S' = S, & \text{otherwise} \end{cases} \tag{4.15}$$

We shall introduce symbols

$$[x,y] \in \mathbb{Z}_\ell$$

for all $x \in S'$ and all $y \in J(\mathbb{Q})$ such that $y_p \in U(\mathbb{Q}_p)$ when $x = p \in S$, and $y_2 \in U(\mathbb{Q}_2)$ when $x = -1$. For this we shall use the elements w_x of (4.14). If first $p \in S$, $p \equiv 1 (\mathrm{mod}\ \ell)$ then

$$w_p^{[p,y]} \equiv y_p (\mathrm{mod}\ p) \tag{4.16a}$$

Next if $p = \ell \neq 2$, $p \in S$, let y'_p be the $p - 1$-st root of unity with $y_p \equiv y'_p (\mathrm{mod}\ p)$. Then

$$y'_p\, w_p^{[p,y]} = y_p. \tag{4.16b}$$

Finally if $p = 2 \in S$ then

$$w_2^{[2,y]} (-1)_2^{[-1,y]} = y_2. \tag{4.16c}$$

Of course $[-1,y]$, and for $p \equiv 1 (\mathrm{mod}\ \ell)$ also $[p,y]$ is not uniquely determined by those equations, but the ambiguity will be seen to be irrelevant.

<u>4.11 Theorem</u>. <u>The group</u> $\Lambda = \Lambda[S]$ <u>has a minimal generating set</u> $\{\tau_x\}$ (<u>all</u> $x \in S'$), <u>satisfying</u>

$$\tau_p^{ab} = (E/\mathbb{Q}, w_p) \quad \underline{\text{for all}} \quad p \in S,$$

and if $2 \in S$

$$\tau_{-1}^{ab} = (E/\mathbb{Q},(-1)_2).$$

A complete set of defining relations of Λ *as a pro* ℓ*-group of class two is*:

$$\tau_p^{p-1} = \prod_{\substack{x \in S' \\ x \neq p}} (\tau_x,\tau_p)^{[x,p]} \quad \text{if} \quad p \in S,\ p \equiv 1 (\text{mod } \ell),$$

$$\tau_{-1}^2 = \prod_{\substack{q \in S \\ q \neq 2}} (\tau_q,\tau_2)^{[q,2]} \quad \text{if} \quad 2 \in S.$$

The inertia group $\Lambda_{(p),0}$, *for* $p \in S$, *is generated modulo* $\Lambda_{(p),0} \cap \Lambda^{(1)}$ *by* τ_p *if* $p \neq 2$, *and by* τ_2 *and* τ_{-1} *if* $p = 2$.

The group $\Lambda_{(p),0} \cap \Lambda^{(1)}$, *for* $p \in S$, *is generated by*

$$\prod_{\substack{x \in S' \\ x \neq p}} (\tau_x,\tau_p)^{[x,p]} \quad \text{if} \quad p \in S,\ p \neq 2,$$

and, for $2 \in S$, *by*

$$\prod_{\substack{q \in S \\ q \neq 2}} (\tau_q,\tau_2)^{[q,2]}, \quad \prod_{\substack{q \in S \\ q \neq 2}} (\tau_q,\tau_{-1})^{[q,2]}, \quad (\tau_2,\tau_{-1}).$$

Proof. The form of the minimal generating set follows from Theorem 4.10, and the same is true for the fact that a complete set of defining relations are obtained by expressing τ_p^{p-1} (for $p \equiv 1(\text{mod } \ell)$) and τ_{-1}^2 (for $2 \in S$) in terms of commutators. To get the actual form of the relations we can use the local Theorems 4.6 and 4.8, i.e., work within the images $\Lambda_{(p)}$ in Λ of the local Galois groups. Let in particular, for each $p \in S$, $\omega_p \in \Lambda$ be an element in $\Lambda_{(p)}$, so that

$$\omega_p^{ab} = (E/\mathbb{Q},p_p). \tag{4.17}$$

Thus ω_p acts as Frobenius at p on the maximal extension of $\mathbb{Q}$ in E in which p is non-ramified. By Theorem 4.8, we have

$$\tau_p^{p-1} = (\tau_p, \omega_p), \text{ for } p \in S, \ p \equiv 1 (\text{mod } \ell). \tag{4.18}$$

By Theorem 4.6, for $F = \mathbb{Q}_2$,

$$\tau_{-1}^2 = (\tau_2, \omega_2), \quad \text{if} \quad 2 \in S. \tag{4.19}$$

But, for all $p \in S$,

$$\prod_{q \in S} (E/\mathbb{Q}, p_q) = 1 \quad (\text{product rule}).$$

Let $q \neq p$. If $q \neq 2$, then $(E/\mathbb{Q}, p_q) = (\tau_q^{[q,p]})^{ab}$, while if $q = 2$, then $(E/\mathbb{Q}, p_2) = (\tau_2^{[2,p]}\tau_{-1}^{[-1,p]})^{ab}$. Hence

$$\omega_p^{ab} = \prod_{x \in S'}{}^{*} (\tau_x^{ab})^{-[x,p]}, \tag{4.20}$$

(product over all $x \neq p$ if p is odd, all $x \neq 2, -1$ if $p = 2$). The relations in the theorem now follow from (4.18), (4.19) and (4.20).

The results on $\Lambda_{(p),0} \bmod \Lambda^{(1)}$ are trivial. Those on $\Lambda_{(p),0} \cap \Lambda^{(1)}$ follow again from (4.20) and the corresponding local results. (In the local case the commutator subgroup lies in the inertia group!) Thus for p odd, $\Lambda_{(p),0} \cap \Lambda^{(1)}$ is generated by (τ_p, ω_p), and for $p = 2$ by (τ_2, ω_2), (τ_{-1}, ω_2) and (τ_2, τ_{-1}).

<u>Remark</u>. Although we did not demand this initially we can of course postulate that the τ_x themselves lie in the appropriate inertia groups.

To complete the arithmetic description of Λ it only remains to show how the image $\Lambda_{(\infty)}$ of the Galois group at infinity is embedded in Λ. Here of course we take $\ell = 2$. We shall also assume that $2 \in S$; if not one just omits the relevant factors.

<u>4.12 Theorem</u>. <u>Let</u>

$$\tau_\infty = \tau_{-1} \cdot \prod_{\substack{p \in S \\ p \neq 2}} \tau_p^{p-1/2},$$

<u>the product being taken in some order of the primes</u> p, <u>e.g.</u>, <u>the ordinary order</u>

of magnitude. *Then* $\Lambda_{(\infty)}$ *is generated by an element* $\tau_\infty \delta_\infty$, $\delta_\infty \in \Lambda^{(1)}$, *where*

$$\delta_\infty^2 = \prod_{p<q} (\tau_q, \tau_p)^{a(q,p)} \prod_p [(\tau_2, \tau_p)^{a(2,p)} (\tau_{-1}, \tau_p)^{a(-1,p)}],$$

with

$$a(p,q) = [p,q] - [q,p] - ((p - 1)/2)\cdot((q - 1)/2),$$

$$a(2,p) = [p,q] - [2,p],$$

$$a(-1,p) = (p - 1)/2 - [-1,p].$$

In the first product factor for δ_∞^2, p, q *run over all pairs of odd primes in* S *with* $p < q$, *in the second product* p *runs over the odd primes* p *in* S.

Corollary. $a(p,q) \equiv a(2,p) \equiv a(-1,p) \equiv 0 \pmod 2$ (Quadratic reciprocity law!)

Proof of the theorem. The image of $\Lambda_{(\infty)}$ in Λ^{ab} is generated by

$$(E/\mathbb{Q},(-1)_\infty), \text{ which is } = \prod_{p \in S} (E/\mathbb{Q},(-1)_p) = \tau_{-1}^{ab} \prod_{\substack{p \in S \\ p \neq 2}} (\tau_p^{ab})^{p-1/2},$$

by the product formula. Thus indeed $\Lambda_{(\infty)}$ is generated by an element $\tau_\infty \delta_\infty$, with $\delta_\infty \in \Lambda^{(1)}$. But $(\tau_\infty \delta_\infty)^2 = 1$, i.e., $\delta_\infty^2 = \tau_\infty^{-2}$. On evaluating τ_∞^{-2} we get exactly the right-hand side of the given equation, using the formulae of Theorem 4.11.

§5. More on Class Groups

Throughout ℓ is a fixed prime and we consider Galois extensions of $\mathbb{Q}$ which are of ℓ-power degree - unless otherwise mentioned. We shall briefly outline the contents of this section in terms of the narrow class group - analogous results will be obtained also for the ordinary class group. We consider a field M, Abelian of ℓ-power degree over $\mathbb{Q}$, which is its own narrow genus field, i.e., $M = G_+(M/\mathbb{Q})$. We shall then determine the Galois group Γ of the narrow central class field $Z_+(M/\mathbb{Q})$ over $\mathbb{Q}$, by generators and relations - entirely in rational terms, more precisely in terms of the finite primes p which are ramified and of their mutual congruence behavior. If, say, K is an Abelian extension of $\mathbb{Q}$ of ℓ-power degree and we take $M = G_+(K/\mathbb{Q})$, and let K be the fixed field of the subgroup Δ of Γ then we get an isomorphism

$$\mathfrak{Cl}_+(K)/\mathfrak{H}_+(K/\mathbb{Q})A(\Gamma/\Delta) \simeq \Delta/(\Delta,\Delta) \tag{5.1}$$

and the group on the right is explicity described. Certain subquotients of this group are of special interest. Thus

$$\mathfrak{H}_+(K/\mathbb{Q})/\mathfrak{H}_+(K/\mathbb{Q})A(\Gamma/\Delta) \simeq (\Gamma,\Gamma)/(\Delta,\Delta), \tag{5.2}$$

and the group on the left is $= 1$ if, and only if, the map from the narrow ℓ-class group of K onto the narrow genus group is an isomorphism, or in other words if, and only if, the narrow genus number is prime to ℓ. For a cyclic field K we shall give an explicit reational criterion for this to happen. On the other hand, by examining the group (Γ,Γ) itself we shall derive a rational description of all Abelian fields of ℓ-power degree whose narrow class number is prime to ℓ (and similarly for the ordinary class number). In fact any problem which can be reformulated in the context of central class fields over $\mathbb{Q}$, can be dealt with in rational terms. Thus for instance, a quotient of the group in (5.2) is

$$\mathfrak{H}_+(K/\mathbb{Q})/\mathfrak{Cl}_+(K)A(\Gamma/\Delta) \simeq (\Gamma,\Gamma)/(\Delta,\Gamma) \tag{5.3}$$

i.e., the number knot $\mathfrak{N}(K/\mathbb{Q})$ (see Theorem 3.11). Details here may be left as an exercise.

Let then $M = G_+(M/\mathbb{Q})$ be an Abelian ℓ-extension of $\mathbb{Q}$. By Theorem 2.15, $\Phi(M/\mathbb{Q})$ has an independent generating set consisting of characters which are ramified at only one finite rational prime divisor, (and possibly at ∞). (We call elements $\{x_i\}$ of a finite Abelian group independent if $\Sigma\, n_i x_i = 0$ implies $n_i x_i = 0$, all i, and if all x_i are different from 0). For each finite prime p, let $\Phi_{(p)}$ be the group of rational idele class characters of ℓ-power order, which are non-ramified at all finite primes other than p. This group is trivial if $p \not\equiv 1 (\mathrm{mod}\ \ell)$ and $p \neq \ell$, so we exclude such primes from consideration. If $p \neq 2$ then every finite subgroup of $\Phi_{(p)}$ is cyclic. For $p = 2$, $\Phi_{(2)} = \Phi'_{(2)} \times \langle\theta\rangle$, where $\theta(u_2) = 1$, if $u_2 \in U(\mathbb{Q}_2)$, $u_2 \equiv 1 (\mathrm{mod}\ 4)$, and $\theta((-1)_2) = -1$. Also $\Phi'_{(2)}$ consists of all characters ϕ' with $\phi'((-1)_2) = 1$. Every finite subgroup of $\Phi'_{(2)}$ is cyclic. Let then throughout S be the set of finite rational prime divisors ramified in M. M is completely determined by a set of independent generators of $\Phi(M/\mathbb{Q})$, and lying in some $\Phi_{(p)}$. Such a set has the following form:

(i) For each $p \in S$, $p \neq 2$, a character $\phi_{(p)} \in \Phi_{(p)}$ of order ℓ^{ν_p}, $(\nu_p \geq 1)$. (If $p \equiv 1 (\mathrm{mod}\ \ell)$ this implies that $p \equiv 1 (\mathrm{mod}\ \ell^{\nu_p})$).

(ii) If $2 \in S$,

either: θ and a character $\phi_{(2)} \in \Phi'_{(2)}$ of order 2^{ν_2} $(\nu_2 \geq 1)$ (case (A)),

or: $\theta\phi_{(2)}$, ($\phi_{(2)} \in \Phi'_{(2)}$ of order 2λ, $\lambda \neq 0$) (case (B)),

or: θ (case (C)),

or: a character $\phi_{(2)} \in \Phi'_{(2)}$ of order 2^{ν_2} $(\nu_2 \geq 1)$ (case (D)).

(5.4)

These cases will be referred to repeatedly in the sequel.

Let Λ be again the Galois group of the maximal ℓ-extension of $\mathbb{Q}$, of class two, in which all finite primes outside S are non-ramified. M and $Z_+(M/\mathbb{Q})$ are subfields of this field, thus the fixed fields of groups Θ, and Σ respectively. Then $\Theta \supset \Lambda^{(1)}$ and, modulo $\Lambda^{(1)}$, the group Θ has a generating set in terms of the inertia generators τ_x of Λ (cf. Theorem 4.11) as follows (with the ν_p and λ as in (5.4)):

$$\begin{array}{lll} \text{For each odd } p \in S, \tau_p^{\ell^{\nu_p}}. & & \\ \text{If } 2 \in S, & & \\ \text{either } \tau_2^{2^{\nu_2}} & \text{(case (A)),} & \\ \text{or } \tau_2^{\lambda}\,\tau_{-1} & \text{(case (B)),} & \\ \text{or } \tau_2 & \text{(case (C)),} & \\ \text{or } \tau_{-1}, \tau_2^{2^{\nu_2}} & \text{(case (D)).} & \end{array} \tag{5.5}$$

We now turn our attention to the subgroup Σ of Λ belonging to $Z_+(M/\mathbb{Q})$. Clearly

$$\Sigma = (\Theta,\Lambda) \prod_{p \in S} (\Lambda_{(p),0} \cap \Theta). \tag{5.6}$$

The generators of Θ in list (5.5) lie in inertia groups, hence they will lie in Σ, and so will their commutators with other elements. Moreover the elements enumerated in the last part of Theorem 4.11 will belong to $\Lambda_{(p),0} \cap \Theta$, hence to Σ. Conversely the fixed field of all these elements will be central and non-ramified over M, hence in $Z_+(M/\mathbb{Q})$. We thus have here a complete generating set of Σ.

We shall want to give the structure of the Galois group of $Z_+(M/\mathbb{Q})$ in terms of generators and relations. We shall therefore use (5.5) to eliminate some generators of Λ/Θ. Indeed the inertia group of 2 is generated by the image of τ_2 in cases (B) and (D) and by that of τ_{-1} in case (C). Modulo (Θ,Λ) we can then also eliminate some of the generators of $\Lambda_{(2),0} \cap \Lambda^{(1)}$ in the list of Theorem 4.11, namely (τ_2,τ_{-1}) in all cases except (A), and $\prod_q (\tau_q,\tau_y)^{[q,2]}$ for $y = 2$ in case (C), and for $y = -1$ in cases (B) and (D). Moreover modulo (Θ,Λ) we can simplify the product $\prod_x (\tau_x,\tau_p)^{[x,p]}$, by omitting (τ_y,τ_p) with $y = 2$ in case (C), with $y = -1$ in case (D), and by replacing

$$(\tau_2,\tau_p)^{[2,p]}(\tau_{-1},\tau_p)^{[-1,p]} \quad \text{by} \quad (\tau_2,\tau_p)^{[2,p]-\lambda[-1,p]}$$

in case (B).

Now recall definitions (4.14)-(4.16) and Theorem 4.11. Summing up, we get

5.1 Theorem. *Let* M *be an Abelian field of* ℓ-*power degree over* $\mathbb{Q}$, *with* $M = G_+(M/\mathbb{Q})$, *and given by* (5.4). *Then*

$$\Gamma = \mathrm{Gal}(Z_+(M/\mathbb{Q})/\mathbb{Q})$$

has a generating set of elements γ_x, *as follows.*

For each $p \in S$, $p \neq 2$ *an element* γ_p *with* $\gamma_p^{ab} = (M/\mathbb{Q},w_p)$

If $2 \in S$

either elements γ_2,γ_{-1} *with* $\gamma_2^{ab} = (M/\mathbb{Q},w_2),\gamma_{-1}^{ab} = (M/\mathbb{Q},(-1)_2)$ (*case* (A)),

or an element γ_2 *with* $\gamma_2^{ab} = (M/\mathbb{Q},w_2)$ (*cases* (B) *and* (D)),

or an element γ_{-1} *with* $\gamma_{-1}^{ab} = (M/\mathbb{Q},(-1)_2)$ (*case* (C)).

Γ *as an* ℓ-*group of class two is then determined by the following relations:*

(i) $\gamma_p^{\ell^{\nu_p}} = 1$ *for all odd* $p \in S$,

and if $2 \in S$,

$\gamma_{-1}^2 = 1$ *in cases* (A), (C)

$\gamma_2^{2^{\nu_2}} = 1$ *in cases* (A), (D),

$\gamma_2^{2\lambda} = 1$ *in case* (B).

(ii) *for all odd* $p \in S$,

$$\prod_{\substack{q\in S\\ q\neq p,2}} (\gamma_q,\gamma_p)^{[q,p].\sigma} = 1,$$

where $\sigma = 1$ if $2 \notin S$, and if $2 \in S$

$\sigma = (\gamma_2,\gamma_p)^{[2,p]}(\gamma_{-1},\gamma_p)^{[-1,p]}$ (case (A)),

$\sigma = (\gamma_2,\gamma_p)^{[2,p]-\lambda[-1,p]}$ (case (B)),

$\sigma = (\gamma_{-1},\gamma_p)^{[-1,p]}$ (case (C)),

$\sigma = (\gamma_2,\gamma_p)^{[2,p]}$ (case (D)).

(iii) If $2 \in S$

$(\gamma_2,\gamma_{-1}) = 1$ in case (A)

$\prod_{\substack{p\in S\\ p\neq 2}} (\gamma_p,\gamma_2)^{[p,2]} = 1$ in cases (A), (B), (D),

$\prod_{\substack{p\in S\\ p\neq 2}} (\gamma_p,\gamma_{-1})^{[p,2]} = 1$ in cases (A) and (C).

We may view the commutators (γ,δ) as images of the elements $\gamma^{ab} \wedge \delta^{ab}$ of $\Gamma^{ab} \wedge {}^{ab}$. In this sense we get the

Corollary 1. $\mathrm{Gal}(Z_+(M/\mathbb{Q})/M) = \Gamma^{(1)}$ is the exterior power $\Gamma^{ab} \wedge \Gamma^{ab}$ (with $\Gamma^{ab} = \mathrm{Gal}(M/\mathbb{Q})$) modulo the relations (ii) and (iii) in Theorem 5.1.

Of course,

$$\Gamma^{(1)} \cong \mathfrak{Cl}_+(M)/\mathfrak{Cl}_+(M)A(\Gamma^{ab}). \tag{5.7}$$

Using the given generators and relations, we can describe the structure of this group in terms of a relation matrix over $\mathbb{Z}_\ell$. We shall content ourselves to reduce mod ℓ, i.e., to work over $\mathbb{F}_\ell$. Let $R(\Gamma^{ab})$ be the ideal $\ell(\mathbb{Z}\Gamma^{ab}) + A(\Gamma^{ab})$ of $\mathbb{Z}(\Gamma^{ab})$. Thus

$$\Gamma^{(1)}/\Gamma^{(1)\ell} \cong \mathfrak{Cl}_+(M)/\mathfrak{Cl}_+(M)R(\Gamma^{ab}) \tag{5.8}$$

This is an $\mathbb{F}_\ell$-vector space. Let now S^* be the set of subscripts x of generators γ_x, listed in Theorem 5.1. Thus S^* consists of all odd $p \in S$, together with either -1 and 2, or -1, or 2, in the various cases for $2 \in S$.

We define an $\mathbb{F}_\ell$-matrix $M_+(M)$. Its column indices are pairs $(x,y) \in S^* \times S^*$ with $x \neq y$, precisely one of (x,y) and (y,x) appearing. Equivalently the columns are in biunique correspondence with a set $\gamma_x^{ab} \wedge \gamma_y^{ab}$ of independent generators of $\Gamma^{ab} \wedge \Gamma^{ab}$. The rows correspond to the relations (ii) and (iii) in Theorem 5.1, the entry in column (x,y) being the exponent of (γ_x,γ_y) in that relation, reduced mod ℓ. For any Abelian field K of ℓ-power degree, let $n(K)$ be the number of independent generators of $\mathrm{Gal}(K/\mathbb{Q})$, i.e., the $\mathbb{F}_\ell$-dimension of $\Phi(K/\mathbb{Q})/\Phi(K/\mathbb{Q})^\ell$. We then have

Corollary 2. $\dim_{\mathbb{F}_\ell}(\mathcal{C}l_+(M)/\mathcal{C}l_+(M)R(\Gamma^{ab})) = \binom{n(M)}{2} - \mathrm{rank}\,(M_+(M))$. For, $\dim_{\mathbb{F}_\ell}(\Gamma^{ab} \wedge \Gamma^{ab}) = \binom{n(M)}{2}$.

We now get a criterion for the narrow class number $h_+(K)$ of an Abelian field of ℓ-power degree to be prime to ℓ. If we want to number the primes in S and if $2 \in S$, we shall make the convention that $p_1 = 2$. We shall view the exponents $[x,y]$ now as elements of $\mathbb{F}_\ell$.

5.2 Theorem. (I) The narrow class number $h_+(M)$ of an Abelian field M of ℓ-power degree is prime to ℓ if, and only if, firstly $M = G_+(M/\mathbb{Q})$ and secondly

$$\mathrm{rank}(M_+(M)) = \binom{n(M)}{2}.$$

(II) Let $M = G_+(M/\mathbb{Q})$. Then $h_+(M)$ is prime to ℓ in exactly the following cases

(a) $n(M) = 1$,

(b) $n(M) = 2$ and

either (i) 2 is the only ramified finite prime (case (A) for $n(M) = 2$) or (ii) S consists of two primes p_1, p_2 and further either $[p_2,p_1] \neq 0$, or (if $2 \notin S$, or $2 \in S$ and we have case (D)) $[p_1,p_2] \neq 0$, or $[-1,p_2] \neq 0$ ($2 \in S$, case (C)), or $[2,p_2] - \lambda[-1,p_2] \neq 0$ ($2 \in S$, case (B)).

(c) $n(M) = 3$ and

<u>either</u> (i) S <u>consists of</u> 2 <u>and an odd prime</u> p_2 (<u>i.e.</u>, <u>we are in case</u> (A)) <u>and</u> $[p_2,2] \neq 0$, <u>i.e.</u>, $(2/p_2) = -1$ <u>or</u> (ii) S <u>consists of three distinct primes and</u>

$$\det\begin{pmatrix} -[p_2,p_1], & 0 & [p_3,p_1] \\ a(p_2) & -[p_3,p_2] & 0 \\ 0 & [p_2,p_3] & -a(p_3) \end{pmatrix} \neq 0$$

<u>where for</u> $p \neq p_1$,

$a(p) = [p_1,p]$ <u>if</u> p_1 <u>is odd or if</u> $p_1 = 2$ <u>in case</u> (D),

$a(p) = [-1,p]$ <u>if</u> $p_1 = 2$, <u>in case</u> (C)

$a(p) = [2,p] - \lambda[-1,p]$ <u>if</u> $p_1 = 2$, <u>in case</u> (B).

<u>If</u> $n(M) > 3$ <u>then</u> $\ell | h_+(M)$.

The conditions given in the theorem solely involve the group $\Phi(M/\mathbb{Q})$ and the mutual congruence behavior of the ramified primes. For instance, if $n(M) = 2$ and, say, S consists of two odd primes, then $h_+(M)$ will be odd precisely if at least one of the primes is not an ℓ-th power residue modulo the other.

In the case $\ell = 2$ the quadratic reciprocity law implies a tightening of the conditions. We shall give some examples for $n(M) = 3$, assuming that S contains three primes. If either all three are odd, or if $p_1 = 2$ and we have case (D), the criterion for $h_+(M)$ to be odd is

$$[p_1,p_2][p_2,p_3][p_3,p_1] - [p_2,p_1][p_3,p_2][p_1,p_3] \neq 0. \tag{5.9}$$

Thus one term, say the first, is 0, and so the second term is 1. This implies $[p_2,p_1] = [p_3,p_2] = [p_1,p_3] = 1$. Thus one of the factors in the first term is 0. If all p_j are odd, we may, without loss of generality, assume that it is $[p_2,p_3]$. If $p_1 = 2$ then by the quadratic reciprocity law $[p_1,p_2] = [p_3,p_1] = 1$, so we must in any case have $[p_2,p_3] = 0$. As $[p_3,p_2] = 1$ this amounts to saying that $(p_2-1)/2\cdot(p_3-1)/2 = 1$ in $\mathbb{F}_2$. We sum up

Addendum I to Theorem 5.2. *Let* $n(M) = 3$, $\ell = 2$, *assume that either* S *contains* 3 *odd primes, or that* $p_1 = 2$, *and we are in case* (D). *A necessary and sufficient condition for* $h_+(M)$ *to be odd is that, with appropriate numbering of the primes*

$$\left(\frac{p_1}{p_2}\right) = \left(\frac{p_2}{p_3}\right) = \left(\frac{p_3}{p_1}\right) = -1$$

(*the last equation to be replaced by* $\left(\frac{2}{p_3}\right) = -1$ *if* $p_1 = 2$) *and that* $p_2 \equiv p_3 \equiv -1 \pmod 4$.

Next we show

Addendum II to Theorem 5.2. *Let* $n(M) = 3$. *Let* S *contain the prime* 2, *and two odd primes* p_2, p_3 *and suppose that we have case* (C). *A necessary and sufficient condition for* $h_+(M)$ *to be odd is that for some numbering of the primes*

$$\left(\frac{-1}{p_2}\right) = \left(\frac{p_3}{p_2}\right) = \left(\frac{2}{p_3}\right) = -1.$$

Proof. Our condition is now that

$$[-1,p_2][p_2,p_3][p_3,2] - [-1,p_3][p_3,p_2][p_2,2] \neq 0.$$

Without loss of generality we can assume that $[-1,p_2] = [p_2,p_3] = [p_3,2] = 1$, which is the stated condition. Then by the quadratic reciprocity law either $[-1,p_3] = 0$ or $[p_3,p_2] = 0$. Hence the result.

One can deal similarly with the case (B). If $\lambda \equiv 0 \pmod 2$ then this has the same features as case (D).

Remark. If ℓ is odd, or if M is complex, then $h_+(M)$ is prime to ℓ precisely if the ordinary class number $h(M)$ is prime to ℓ. If however M is real, then $h_+(M)$ is odd if, and only if, firstly $h(M)$ is odd and secondly all signatures are taken by global units (by the exactness of the sequence (1.19)).

Corollary 1. (Weber's Theorem). *Any subfield of the field of* 2^n*-th roots*

<u>of unity over</u> $\mathbb{Q}$, <u>has odd narrow class number</u> h_+, (<u>for all</u> n).

For, every such field M coincides with $G_+(M/\mathbb{Q})$. Now apply case (b) (i) in Theorem 5.2 (II).

Let $t(K)$ be the number of finite primes divisors of $\mathbb{Q}$ ramified in K.

<u>Corollary 2</u>. <u>Let</u> K <u>be an Abelian field of</u> ℓ-<u>power degree</u>. <u>If</u> $n(K) > 3$ <u>or if</u> $t(K) > 3$ <u>then</u> $\ell | h_+(K)$.

<u>Proof</u>. We may suppose that $K = G_+(K/\mathbb{Q})$. For such a field $n(K) \geq t(K)$. Now apply the theorem.

<u>Corollary 3</u>. <u>Let</u> $L = \mathbb{Q}(e^{2\pi i/m})$, m <u>odd or</u> $m \equiv 0 \pmod 4$. <u>If</u> m <u>has more than three prime factors then</u> $h(L)$ <u>is even</u>. <u>If</u> m <u>has more than three prime factors which are either</u> $\equiv 1 \pmod \ell$ <u>or</u> $= \ell$, <u>then</u> $\ell | h(L)$.

<u>Proof</u>. Let M be the maximal ℓ-extension of $\mathbb{Q}$ in L. Then $M = G_+(M/\mathbb{Q})$ and by the Theorem $\ell | h_+(M)$. As L/M is totally ramified, and L is complex this implies that $\ell | h(L)$.

<u>Proof of Theorem 5.2</u>. By the Corollary to Proposition 2.7, the degree $[G_+(M/\mathbb{Q}) : M]$ is a power of ℓ, if $M/\mathbb{Q}$ is Galois of ℓ-power degree. The degree divides $h_+(M)$. Thus we must assume that $M = G_+(M/\mathbb{Q})$. By Lemma 3.9, $h_+(M)$ is prime to ℓ if, and only if, $\mathfrak{Cl}_+(M)/\mathfrak{Cl}_+(M)A(\Gamma^{ab}) = 1$, or equivalently $\mathfrak{Cl}_+(M)/\mathfrak{Cl}_+(M)R(\Gamma^{ab}) = 1$. Now apply Corollary 2 to Theorem 5.1, to get the condition under (I).

For (II) we first note that the number of rows of $M_+(M)$ is at most $n(M) + 1$. Hence $\operatorname{rank}(M_+(M)) \leq n(M) + 1$. But for $n > 3$, also $\binom{n}{2} > n + 1$. Hence if $n(M) > 3$ then $\ell | h_+(M)$.

The case $n(M) = 1$ goes back to the genus theory of cyclic fields - see Corollary 1 to Theorem 2.16. Alternatively in this case

$$\binom{n(M)}{2} = 0 = \operatorname{rank} M_+(M).$$

All the other cases are obtained by writing down the matrix $M_+(M)$. The least obvious is the case when $n(M) = 3$ and S contains only two primes, 2 and an odd prime p. The matrix $M_+(M)$ has three columns with indices $\tau_2 \wedge \tau_{-1}$, $\tau_2 \wedge \tau_p$, $\tau_{-1} \wedge \tau_p$ and has rows as indicated, giving the relations:

$$\begin{array}{l|}
(\tau_2,\tau_{-1}) = 1 \\
(\tau_2,\tau_p)^{[2,p]}(\tau_{-1},\tau_p)^{[-1,p]} = 1 \\
(\tau_p,\tau_2)^{[p,2]} = 1 \\
(\tau_p,\tau_{-1})^{[p,2]} = 1
\end{array}
\begin{pmatrix} 1 & 0 & 0 \\ 0 & [2,p] & [-1,p] \\ 0 & [p,2] & 0 \\ 0 & 0 & [p,2] \end{pmatrix}$$

As $[2,p] = [p,2]$, the result follows.

We now turn to the question of the divisibility by ℓ of the order $g_+(L)$ of theprincipal genus $\mathfrak{H}_+(L/\mathbb{Q})$, when $L/\mathbb{Q}$ is cyclic of degree a power of ℓ. By Corollary 1 to Theorem 2.16, this question is only relevant if the number $t(L)$ of finite rational primes ramified in L is at least 2, and so $L \neq G_+(L/\mathbb{Q})$. We write $M = G_+(L/\mathbb{Q})$. Then M is a field as considered in the earlier part of this section, in particular in Theorems 5.1 and 5.2. We shall continue to employ the notation introduced earlier in connection with the study of M. In particular $\Gamma = \mathrm{Gal}(Z_+(M/\mathbb{Q})/\mathbb{Q})$, and for $2 \in S$ we need the case distinctions. However, case (A) cannot now occur. For, as $\Phi(L/\mathbb{Q})$ is cyclic, so is $\Phi(L/\mathbb{Q})_{(p)}$, which by Theorem 2.14 coincides with $\Phi(M/\mathbb{Q})_{(p)}$.

Let now p_j $(j = 1,\dots,t)$ be the finite rational primes which are ramified in L, i.e., in M. We shall write $\gamma_{p_i} = \gamma_i$. For one prime at least the ramification index in L is the degree $[L : \mathbb{Q}]$. Let p_1 be such. Let now Δ be the subgroup of Γ, fixing L. Then mod $\Gamma^{(1)}$, Δ has an independent generating set of form $\{\gamma_1^{-r_j}\gamma_j\}$ $(j = 2,\dots,t)$. For convenience we write $r_1 = 1$. We define a matrix $S_+(L)$ with coefficients in $\mathbb{F}_\ell$ as follows. Its column indices are $j = 2,\dots,t$, and its row indices $k = 1,\dots,t$. The entry in the place k,k $(k \geq 2)$ is

$$\sum_{j=1,j\neq k}^{t} r_j[p_j,p_k]$$

and the entry in the place k,j with $k \neq j$ is

$$-r_k[p_j,p_k],$$

where the r_j and the $[p_j,p_k]$ are viewed as elements of $\mathbb{F}_\ell$. This is under the hypothesis that either all p_j are odd, or that 2 occurs and we are in case (D). If 2 occurs and we have case (C) or case (B), then throughout the definition of $S_+(L)$, $[2,p_k]$ (p_k odd) has to be replaced by $[-1,p_k]$, or by $[2,p_k] - \lambda[-1,p_k]$, respectively.

5.3 Theorem. Let $L/\mathbb{Q}$ be cyclic of ℓ-power degree, and p_j ($j = 1,\dots,t$) the finite rational primes ramified in L. Then

$$\dim_{\mathbb{F}_\ell}(\mathfrak{H}_+(L/\mathbb{Q})/[\mathfrak{H}_+(L/\mathbb{Q})^{1-\gamma_1}\mathfrak{H}_+(L/\mathbb{Q})^\ell]) = t - 1 - \mathrm{rank}(S_+(L)).$$

In particular the order $g_+(L)$ of the narrow principal genus $\mathfrak{H}_+(L/\mathbb{Q})$ is prime to ℓ if, and only if, $\mathrm{rank}(S_+(L)) = t - 1$.

Remark. As the proof will show, the matrix $S_+(L)$, viewed as a matrix over $\mathbb{Z}_\ell$, and extended by a matrix whose entries are the orders of the γ_j^{ab} ($j > 1$) in Γ^{ab}, is actually the relation matrix of $\mathfrak{H}_+(L/\mathbb{Q})/\mathfrak{H}_+(L/\mathbb{Q})^{1-\gamma_1}$.

Proof of Theorem 5.3. We know that

$$\mathfrak{H}_+(L/\mathbb{Q})/\mathfrak{H}_+(L/\mathbb{Q})^{1-\gamma_1} \cong (\Gamma,\Gamma)/(\Delta,\Delta). \tag{5.10}$$

Let for the moment $\Psi = \mathrm{Ker}[(\Gamma^{ab}{}_\wedge\,\Gamma^{ab}) \to (\Gamma,\Gamma)]$. Then

$$(\Gamma,\Gamma)/(\Delta,\Delta) = \mathrm{Cok}[\Psi \to \Gamma^{ab}{}_\wedge\,\Gamma^{ab}/\Delta^{ab}{}_\wedge\,\Delta^{ab}].$$

But $\Gamma^{ab}{}_\wedge\,\Gamma^{ab}/\Delta^{ab}{}_\wedge\,\Delta^{ab}$ is isomorphic to the submodule of $\Gamma^{ab}{}_\wedge\,\Gamma^{ab}$ generated by the independent elements $\gamma_1^{ab}{}_\wedge\,\gamma_j^{ab}$ ($j > 1$), under the map

$$\gamma_i^{ab}{}_\wedge\,\gamma_j^{ab} \longmapsto (\gamma_1^{ab})^{-r_j}{}_\wedge\,\gamma_i^{ab} - (\gamma_1^{ab})^{-r_i}{}_\wedge\,\gamma_j^{ab}.$$

This follows on expanding the congruence

$$[(\gamma_1^{ab})^{-r_i}\gamma_i^{ab}]\wedge[(\gamma_1^{ab})^{-r_j}\gamma_j^{ab}] \equiv 0 (\mathrm{mod}\ \Delta^{ab}{}_\wedge\,\Delta^{ab}).$$

In other words, we get the relations, defining $(\Gamma,\Gamma)/(\Delta,\Delta)$ as a quotient of

$\Gamma^{ab} \wedge \Gamma^{ab}/\Delta^{ab} \wedge \Delta^{ab}$ by substituting

$$(\gamma_i,\gamma_j) \longmapsto (\gamma_1,\gamma_j)^{r_i}(\gamma_1,\gamma_i)^{-r_j}$$

into the relations of Theorem 5.1. By (5.10) this immediately yields the first part of Theorem 5.3. The second part follows from the fact that $\mathfrak{H}_+(L/\mathbb{Q})/\mathfrak{H}_+(L/\mathbb{Q})^{1-\sigma}$ is an ℓ-group.

We now turn to the ordinary absolute class group. When ℓ is odd, then the ℓ-parts of $\mathfrak{Cl}_+(K)$ and $\mathfrak{Cl}(K)$ coincide. Similarly if K is totally complex, then $\mathfrak{Cl}_+(K) = \mathfrak{Cl}(K)$. Thus from now on we take $\ell = 2$, and we shall study real Abelian fields of 2-power degree. But the theory of fields $M = G_+(M/\mathbb{Q})$, as in Theorem 5.1, will still be needed. Let in particular N be a real Abelian field, whose degree is a power of 2, which is its own genus field, i.e.,

$$N = G(N/\mathbb{Q}). \tag{5.11}$$

We put

$$M = G_+(N/\mathbb{Q}). \tag{5.12}$$

Then M may be taken as described by (5.4), and we shall use the notation introduced in that context. In particular S is the set of finite rational primes, which are ramified in M, or equivalently which are ramified in N. By Theorem 2.15, either $N = M$, or M is complex and relative quadratic over N. The former is the case if, and only if, $\phi((-1)_\infty) = 1$ for all $\phi \in \Phi(M/\mathbb{Q})$. To give a condition in terms of Galois groups, let γ_∞ be the image in $\Gamma = \mathrm{Gal}(Z_+(M/\mathbb{Q})/\mathbb{Q})$ of the element τ_∞ of Theorem 4.12, i.e.,

$$\gamma_\infty = \gamma' \cdot \prod_{\substack{p\in S \\ p\neq 2}} \gamma_p^{p-1/2} \quad \text{(product over } p \text{ in a given order)}, \tag{5.13a}$$

where

$$\gamma' = \begin{cases} \gamma_{-1}, & \text{if } 2 \in S, \text{ in case (A) and (C)}, \\ \gamma_2^{-\lambda}, & \text{if } 2 \in S, \text{ in case (B)}, \\ 1, & \text{otherwise} \end{cases} \tag{5.13b}$$

5.4 Lemma. *$N = M$ if, and only if, $\gamma_\infty^{ab} = 1$, i.e., $\gamma_\infty = 1$, i.e., in (5.13), $\gamma' = 1 = \gamma_p^{p-1/2}$ for all odd $p \in S$ (in other words, $p \equiv 1 (\text{mod } 2^{\nu_p+1})$). In particular, if $N = M$ then all odd primes in S are $\equiv 1 \pmod 4$.*

If $N \neq M$ then at least two factors of the product (5.13a) are $\neq 1$.

The last assertion is a consequence of the hypothesis that M/N is non-ramified at all finite primes. The remainder of the Lemma is a consequence of the fact that N is the fixed field of γ_∞^{ab} in $\text{Gal}(M/\mathbb{Q})$, and of the structure theory of the group Γ, as given in Theorem 5.1.

For the next theorem we introduce an element β_∞ of Γ, but only in the case when $N = M$, by the equation

$$\beta_\infty = \prod_{p<q} (\gamma_q,\gamma_p)^{([p,q]-[q,p])/2} \cdot \prod_p (\gamma_2,\gamma_p)^{([p,2]-[2,p])/2}, \qquad (5.14)$$

where the first product runs over all pairs (p,q) of odd primes in S, taken in the given order (the same as in (5.13a)), and the second product runs over the odd primes p in S - but has to be omitted if $2 \notin S$. (Recall that by Lemma 5.4, $2 \in S$ and $N = M$ implies that we are in case (D)). Note that now $p \equiv q \equiv 1 \pmod 4$ so that the exponents in (5.14) are well defined.

5.5 Theorem. *Let N be a real Abelian field of 2-power degree, which is its own genus field. Let M be given by (5.12), and described by (5.4).*

(I) If $N = M$, then $Z(N/\mathbb{Q})$ is the fixed field of an element β'_∞ in (Γ,Γ), with $\beta'_\infty \equiv \beta_\infty (\text{mod}(\Gamma,\Gamma)^2)$.

(II) If $N \neq M$, then $Z(N/\mathbb{Q})$ is the fixed field of $(\Gamma,\langle\gamma_\infty\rangle)$, and a further element $\gamma'_\infty \in \Gamma$, which is $\equiv \gamma_\infty(\text{mod}(\Gamma,\Gamma))$. In particular

$$\text{Gal}(Z(N/\mathbb{Q})/N) \cong (\Gamma,\Gamma)/(\Gamma,\langle\gamma_\infty\rangle).$$

Remark. In both cases the statements can be made more precise, following Theorem 4.12. This however will not be needed.

Proof. Let γ'_∞, and β'_∞ be the images in Γ of the elements $\tau_\infty\delta_\infty$, and δ_∞, respectively, which appeared in Theorem 4.12. Then indeed $\gamma'_\infty = \gamma_\infty\beta'_\infty$, and

$\beta'_\infty \in (\Gamma,\Gamma)$. Clearly $Z(N/\mathbb{Q})$ will be the fixed field of the subgroup Γ_∞ of Γ generated by γ'_∞ and $(<\gamma'_\infty>,\Gamma)$. As $\beta'_\infty \in \mathrm{cent}(\Gamma)$, $(<\gamma'_\infty>,\Gamma) = (<\gamma_\infty>,\Gamma)$. Thus in the case $N \neq M$, the group Γ_∞ has the stated description. Moreover N will be the fixed field of $(\Gamma,\Gamma)\Gamma_\infty$ and clearly

$$(\Gamma,\Gamma)/(<\gamma_\infty>,\Gamma) \simeq (\Gamma,\Gamma)\Gamma_\infty/\Gamma_\infty = \mathrm{Gal}(Z(N/\mathbb{Q})/N).$$

Next in the case $N = M$, we have $\gamma_\infty = 1$, by Lemma 5.4. Hence indeed $Z(N/\mathbb{Q})$ is the fixed field of β'_∞. We have to establish the stated congruence $\mathrm{mod}(\Gamma,\Gamma)^2$. For this we go back to Theorem 4.12, and use the notation introduced there. We know that the exponents $a(x,p)$ are even. So define, in the group Λ of Theorems 4.11 and 4.12,

$$\delta^*_\infty = \prod_{p<q} (\tau_q,\tau_p)^{a(q,p)/2} \prod_p [(\tau_2,\tau_p)^{a(2,p)/2}(\tau_{-1},\tau_p)^{a(-1,p)/2}]. \tag{5.15}$$

Then $\delta_\infty = \delta^*_\infty\delta''_\infty$, where $\delta''_\infty \in (\Lambda,\Lambda)$ and $\delta''^2_\infty = 1$. Now β'_∞ is the image of δ_∞ in Γ, and we denote by β^*_∞ and β''_∞ the images of δ^*_∞, and of δ''_∞ respectively. Thus

$$\beta'_\infty = \beta^*_\infty\beta''_\infty. \tag{5.16}$$

By Lemma 5.4, the orders of the τ_p (p odd) are divisible by 4, and of course τ_2 is of infinite order. Therefore the commutators (τ_q,τ_p), p odd, q odd or $q = 2$, have order divisible by 4. As $\delta''^2_\infty = 1$, it follows, on expanding in terms of commutators of inertia generators, that $\delta''_\infty = \sigma^2(\tau_{-1},\sigma')$, $\sigma \in (\Lambda,\Lambda)$. As the image of (τ_{-1},σ') in Γ is 1, we conclude that

$$\beta''_\infty \in (\Gamma,\Gamma)^2. \tag{5.17}$$

Now β_∞ and β^*_∞ differ by a product of terms

$$(\tau_q,\tau_p)^{c(p,q)};\ c(p,q) = \tfrac{1}{2}((p-1)/2)((q-1)/2),\ p \equiv q \equiv 1 (\mathrm{mod}\ 4).$$

Clearly $c(p,q)$ is even. This in conjunction with (5.16), (5.17) completes the proof.

The last theorem gives us a description of $\mathbb{C}l(N)/\mathbb{C}l(N)A(\Gamma'^{ab})$ (Γ' = $\mathrm{Gal}(Z(N/\mathbb{Q})/\mathbb{Q})$) by generators and relation. Indeed

$$\mathbb{C}l(N)/\mathbb{C}l(N)A(\Gamma'^{ab}) \cong (\Gamma',\Gamma') = \mathrm{Gal}(Z(N/\mathbb{Q})/N). \tag{5.18}$$

We shall obtain a matrix criterion analogous to that in the narrow case. With N and M as in the theorem, we define a matrix $M(N)$ over $\mathbb{F}_2$ (or in the first place over $\mathbb{Z}_2$). Its columns are the same as those of $M_+(M)$, and indeed

$$M(N) = \begin{pmatrix} M_+(M) \\ M'(N) \end{pmatrix} \tag{5.19a}$$

i.e., it is $M_+(M)$ bordered with some extra rows. First, if $N = M$, then $M'(N)$ has exactly one row, the entry in the column q,p being the exponent of (γ_q,γ_p) in the expansion of β_∞, i.e., (cf. (5.14))

$$([p,q] - [q,p])/2. \tag{5.19b}$$

On the other hand, if $N \neq M$, there are $n(M)$ rows (recall that $n(M)$ is the number of independent generator of $\mathrm{Gal}(M/\mathbb{Q})$), the entries in the row with index q, being the exponent of (γ_q,γ_x) in the expansion of (γ_q,γ_∞), with the usual extra convention for the various cases, when $2 \in S$. Thus the q,p entry for p odd will be

$$p - 1/2, \tag{5.19c}$$

and the $q, -1$ entry will be $= 1$.

From Theorem 5.5 we now get the

Corollary.

$$\dim_{\mathbb{F}_2}(\mathbb{C}l(N)/(\mathbb{C}l(N)A(\Gamma'^{ab})\mathbb{C}l(N)^2) = \binom{n(M)}{2} - \mathrm{rank}(M(N)).$$

Remark. Definitions and results can be formally extended to every Abelian field N of 2-power degree, which coincides with its genus field, by setting $M(N) = M_+(N)$ if N is complex.

Going back for the moment to arbitrary ℓ, the entries in the matrices $M_+(M)$, and similarly in $M(N)$, will of course depend on the original choices of the w_p in (4.14). A change in these will amount to a multiplication of the coefficients $[p,x]$ by an ℓ-adic unit. It is obvious, however, that the $\mathbb{F}_\ell$-rank of $M_+(M)$, or of $M(N)$, in the case when $M \neq N$, will not be affected by this, nor even will their elementary divisor, when considered over $\mathbb{Z}_\ell$. This is less obvious in the case of the coefficients (5.19b), although the preceding Corollary implies the invariance of the rank. One can however give a direct argument quite easily. In fact suppose that we have changed w_p (where p may be odd or = 2). This implies that we have, for all q, to replace $[p,q]$ by $u[p,q]$, where $u \in \mathbb{Z}_2^*$. We rewrite

$$(u[p,q] - [q,p])/2 = u([p,q] - (1 + 2a)[q,p])/2,$$

where $1 + 2a = u^{-1}$. The new matrix is thus obtained by first adding a times the row corresponding to the relation $\Pi_q\,(\gamma_q,\gamma_p)^{[q,p]} = 1$ to the last row, and then multiplying the last row by u.

In one case, however, do the coefficients (5.19b) modulo 2 stay invariant and can be described nicely in terms of residue class characters. Let p be an odd prime, $p \equiv 1 \pmod 4$, q an odd prime, $q \equiv 1 \pmod 4$, or $q = 2$, and suppose that $(q/p) = 1$, i.e., $[p,q] \equiv 0 \pmod 2$, hence $[q,p] \equiv 0 \pmod 2$. Then the value of $([p,q] - [q,p])/2$ modulo 2 does not change if we multiply $[p,q]$ by a 2-adic unit. Indeed now

$$\begin{cases} \left(\frac{q}{p}\right)_4 = (-1)^{[p,q]/2}, \text{ for any primitive biquadratic} \\ \text{residue class character } (-)_4 \bmod p. \end{cases} \tag{5.20a}$$

Analogously on interchange if q is odd. If $q = 2$, then our hypothesis implies that $p \equiv 1 \pmod 8$, and we can then put directly

$$\left(\frac{p}{2}\right)_4 = (-1)^{[2,p]/2} = \begin{cases} 1, & \text{if } p \equiv 1 \pmod{16}, \\ -1, & \text{if } p \equiv 9 \pmod{16}. \end{cases} \tag{5.20b}$$

Concluding we get in this case

$$(-1)^{([p,q]-[q,p])/2} = \left(\frac{q}{p}\right)_4 \left(\frac{p}{q}\right)_4 . \tag{5.20c}$$

5.6 Theorem. *Let* N *be a real Abelian field of* 2-*power degree. The class number* h(N) *is odd if, and only if, firstly* $N = G(N/\mathbb{Q})$, *and secondly for* $M = G_+(N/\mathbb{Q})$

$$\binom{n(M)}{2} = \text{rank } \mathcal{M}(N).$$

In particular h(N) *is even, whenever* $t(N) > 4$, *where* t(N) *is the number of finite rational prime divisors ramified in* N. *In the case* N = M, h(N) *will be even, whenever* $n(N) > 3$, *and so whenever* $t(N) > 3$.

Proof. The first part of the theorem follows immediately from the definition of a genus field, from Theorem 5.5, and from the fact that $\mathfrak{Cl}(N)/\mathfrak{Cl}(N)\cdot A(\Gamma'^{ab})$ is a 2-group. The argument is exactly the same as for Theorem 5.2.

Suppose now that $N = G(N/\mathbb{Q})$, but $N \neq M$. The number of rows of $\mathcal{M}_+(M)$ is at most $t(N) + 2$; it is in fact t(N), except when $2 \in S$ and we have case (A). We get $n(M) \leq t(N) + 1$ further rows, from the commutator relations for (γ_∞, σ). But then at most $n(M) - 1 \leq t(N)$ of those are linearly independent. Thus $\text{rank}(\mathcal{M}(N)) \leq 2n(M)$, and - excepting case (A) when $2 \in S$ - even $\text{rank}(\mathcal{M}(N)) \leq 2n(M) - 1$. In that exceptional case $n(M) = t(N) + 1$, otherwise always $n(M) = t(N)$. Substituting, we get

$$\binom{n(M)}{2} > \text{rank } \mathcal{M}(N) \quad \text{for} \quad t(N) > 4.$$

Now suppose that N = M. By Lemma 5.4, $t(N) = n(M)$ in this case. The number of rows of $\mathcal{M}(N)$ is now $n(M) + 1$, and of course $\binom{n}{2} - (n + 1) > 0$ if $n > 3$. This gives the final assertion in the theorem.

We shall mention some explicit special cases of the last theorem, for a real field N.

First suppose that $n(M) = 2$. This implies $t(N) = 2$. For, $t(N) = 1$ would imply that only the prime 2 is ramified, but then either $n(M) = 1$, or

N is complex - both of which we excluded.

Addendum I to Theorem 5.6. *Suppose* $n(M) = 2$, *and* $N \neq M$. *Then* $h(N)$ *is odd if, and only if*

either $2 \in S$, *and we are in case* (C) *or in case* (B) *with* $\lambda \not\equiv 0 \pmod 2$,

or S *contains a prime* $p \equiv -1 \pmod 4$,

or S *consists of two primes* p,q, *with* q *odd, so that*

$$\left(\frac{p}{q}\right) = -1.$$

Suppose now that $n(M) = 2$ *and* $N = M$. *Then* $h(N)$ *is odd if, and only if,*

either $\left(\frac{q}{p}\right) = -1$, $q,p \in S$, p *odd,*

or $\left(\frac{q}{p}\right) = 1$ *but* $\left(\frac{q}{p}\right)_4\left(\frac{p}{q}\right)_4 = -1$.

The details of the proof are laborious, but easy. We leave it to the reader.

Next we look at the case $n(M) = t(N) = 3$. We shall suppose that the primes in S are all odd. The details, when $2 \in S$, are rather similar, and we shall omit them here. Anyone interested can copy our proof.

Addendum II to Theorem 5.6. *Suppose* $n(M) = t(N) = 3$, S *consisting of odd primes* p_j $(j = 1,2,3)$.

Suppose also that $N \neq M$. *Then* $h(N)$ *is odd if, and only if, for some ordering of subscripts*

either $p_3 \equiv -1 \pmod 4$, $p_1 \equiv p_2 \equiv 1 \pmod 4$ *and* $\left(\frac{p_1}{p_2}\right) = -1$,

or $p_3 \equiv p_2 \equiv -1 \pmod 4$, $p_1 \equiv 1 \pmod 4$ *and*

$\left(\frac{p_1}{p_j}\right) = -1$ *for at least one value of* $j = 2,3$,

or $p_1 \equiv p_2 \equiv p_3 \equiv -1 \pmod 4$.

Next suppose that $N = M$, *whence* $p_j \equiv 1 \pmod 4$ *for all* j. *We shall put*

$$c_{ij} = (-1)^{([p_i,p_j]-[p_j,p_i])/2}$$

(this is symmetric, viewed in $\mathbb{F}_2$). Then $h(N)$ is odd if, and only if, for some ordering of subscripts,

$$c_{12} = \left(\frac{p_2}{p_3}\right) = \left(\frac{p_3}{p_1}\right) = -1, \text{ and}$$

$$\text{either } \left(\frac{p_1}{p_2}\right) = 1,$$

$$\text{or } c_{31} = c_{23} = 1,$$

$$\text{or } \left(\frac{p_1}{p_2}\right) = c_{31} = c_{23} = -1.$$

Proof. We take first the case $N \neq M$. Recall that

$$M_+(M) = \begin{pmatrix} [p_2,p_1] & 0 & [p_3,p_1] \\ [p_1,p_2] & [p_3,p_2] & 0 \\ 0 & [p_2,p_3] & [p_1,p_3] \end{pmatrix}.$$

The bordering matrix in (5.19a) is now

$$M'(N) = \begin{pmatrix} p_2 - 1/2 & 0 & p_3 - 1/2 \\ p_1 - 1/2 & p_3 - 1/2 & 0 \\ 0 & p_2 - 1/2 & p_1 - 1/2 \end{pmatrix}.$$

If all p_j are $\equiv 1 \pmod 4$ this is the null matrix, and clearly $M_+(M)$ has at most rank 2, as by the quadratic reciprocity law the sum of the rows is zero.

Next suppose that $p_3 \equiv -1 \pmod 4$, but $p_1 \equiv p_2 \equiv 1 \pmod 4$. Then, removing a row of zeros from $M'(N)$, we are left with the matrix

$$\begin{pmatrix} 0 & 0 & 1 \\ 0 & 1 & 0 \end{pmatrix}.$$

Removing the third row of $M_+(M)$, which depends on the other two, we have "reduced" $M(N)$ to a matrix

$$\begin{pmatrix} [p_2,p_1] & 0 & [p_3,p_1] \\ [p_1,p_2] & [p_3,p_2] & 0 \\ 0 & 0 & 1 \\ 0 & 1 & 0 \end{pmatrix}.$$

If $[p_2,p_1] = 0$, this has rank 2. So we need $[p_2,p_1] = 1$, and this will clearly also suffice.

Similarly we can deal with the case $p_2 \equiv p_3 \equiv -1 \pmod 4$, but $p_1 \equiv 1 \pmod 4$. Finally in the case when all p_j are $\equiv -1 \pmod 4$, if we add the first two rows of $M(N)$ to the third, then add the first three rows both to the fourth and to the fifth, we get a non-singular 3×3 submatrix

$$\begin{pmatrix} 1 & 1 & 1 \\ 0 & 1 & 0 \\ 0 & 0 & 1 \end{pmatrix}.$$

In the case $N = M$, the first three rows are linearly dependent. Removing, say, the top row, and writing $a_{ij} = ([p_i,p_j] - [p_j,p_i])/2$, we get a matrix

$$\begin{pmatrix} [p_1,p_2] & [p_3,p_2] & 0 \\ 0 & [p_2,p_3] & [p_1,p_3] \\ a_{12} & a_{23} & a_{31} \end{pmatrix}$$

Expanding we find that its determinant is

$$[p_1,p_2][p_2,p_3]a_{31} + [p_3,p_2][p_1,p_3]a_{12} + [p_1,p_2][p_1,p_3]a_{23}.$$

Now interpret!

One can also derive a criterion analogous to that given in Theorem 5.3. The procedure should be quite clear, and we shall therefore omit this.

<u>Quadratic fields</u>. Let L be a quadratic field. If $\langle\sigma\rangle = \mathrm{Gal}(L/\mathbb{Q})$, then σ acts on $\mathfrak{Cl}_+(L)$, and on $\mathfrak{Cl}(L)$ by $c \longrightarrow c^{-1}$. Therefore $t(L) - 1$ is the number of invariants of $\mathfrak{Cl}_+(L)$ divisible by 2, and $t(L) - 1 - \mathrm{rank}(S_+(L))$ is the number of invariants of $\mathfrak{Cl}_+(L)$ divisible by 4. Moreover, if L is

real, then the number of even invariants of $\mathfrak{Cl}(L)$ is $t(L) - 1$ or $t(L) - 2$ according to whether $G(L/\mathbb{Q}) =$ or $\neq G_+(L/\mathbb{Q})$. (This is the old distinction between $N = M$, and $N \neq M$).

One can also get information on the norm $N\gamma$ of a fundamental unit of a real quadratic field L. Indeed, by the exactness of the sequence (1.19), $N\gamma = -1$ if, and only if, $\mathfrak{Cl}_+(L) \cong \mathfrak{Cl}(L)$. Thus if $G(L/\mathbb{Q}) \neq G_+(L/\mathbb{Q})$ then $N\gamma = 1$. This is of course a triviality, as the condition amounts to the fact that -1 is not a norm residue. Assume now that -1 is a norm residue, i.e., it is a norm from L, in other words, that $G(L/\mathbb{Q}) = G_+(L/\mathbb{Q})$. Then we can get two results. Firstly if $\mathfrak{H}_+(L/\mathbb{Q})$ has odd order (see criterion of Theorem 5.3) then $\mathfrak{Cl}_+(L) = \mathfrak{Cl}(L)$, hence $N\gamma = -1$. On the other hand, if $\mathfrak{H}_+(L/\mathbb{Q})/\mathfrak{H}_+(L/\mathbb{Q})^{1-\sigma} \neq \mathfrak{H}(L/\mathbb{Q})/\mathfrak{H}(L/\mathbb{Q})^{1-\sigma}$ (see (5.2)) then $N\gamma = 1$.

Finally and most trivial of all, the fact that $h_+(L) = 1$ when $t(L) = 1$ implies that $N\gamma = -1$.

<u>Biquadratic fields</u>. We end up with a description of all biquadratic field

$$L = \mathbb{Q}(\sqrt{d_1}, \sqrt{d_2})$$

with odd discriminant and odd class number. If p is an odd prime, we write

$$p^* = \left(\frac{-1}{p}\right) p.$$

<u>Theorem 5.7</u>. <u>Let</u> L <u>be a biquadratic field</u> (<u>i.e.</u>, <u>a composite of two disjoint quadratic fields</u>). <u>Suppose that</u> L <u>has odd discriminant</u>. <u>Then</u> $h(L)$ <u>is odd if</u>, <u>and only if</u>,

<u>either</u> (a) $L = \mathbb{Q}\left(\sqrt{p_1^*}, \sqrt{p_2^*}\right)$,

p_1, p_2 <u>distinct odd primes and</u> $\left(\frac{p_1}{p_2}\right) = -1$ <u>or</u> $\left(\frac{p_2}{p_1}\right) = -1$ (<u>or both</u>),

<u>or</u> (b) $L = \mathbb{Q}(\sqrt{p_1}, \sqrt{p_2})$,

p_1, p_2 <u>distinct primes</u> $\equiv 1 \pmod 4$, $\left(\frac{p_1}{p_2}\right) = 1$,

<u>and</u> $\left(\frac{p_1}{p_2}\right)_4\left(\frac{p_2}{p_1}\right)_4 = -1,$

<u>or</u> (c) $L = \mathbb{Q}(\sqrt{p_1p_2}, \sqrt{p_2p_3}, \sqrt{p_3p_1})$.

p_j <u>distinct primes</u> $\equiv -1 \pmod 4$,

<u>or</u> (d) $L = \mathbb{Q}(\sqrt{p_1}, \sqrt{p_2p_3})$,

<u>where the</u> p_j <u>are distinct primes</u>, $p_1 \equiv 1$, $p_2 \equiv p_3 \equiv -1 \pmod 4$, $\left(\frac{p_1}{p_j}\right) = -1$, <u>for at least one value of</u> $j = 2,3$.

<u>Proof</u>. We must have $L = G(L/\mathbb{Q})$. If actually $L = G_+(L/\mathbb{Q})$, i.e., in (a), (b) above we use Theorem 5.2 and Addendum I to Theorem 5.6. In the other cases we use Addendum II, observing that now the number of primes $\equiv -1 \pmod 4$ must be at least two.

<u>Exercise</u>. Establish a similar list for biquadratic fields with even discriminant!

§6. Some Remarks on History and Literature

The study of central extensions of number fields was started by A. Scholz (1936, 1940) who was already aware of the connection with the norm problem. Another early contribution is due to L. Rédei (partly in collaboration with H. Reichardt) who obtained results on class groups in quadratic fields outside the range of classical genus theory, by what are in effect central extension methods (1933, 1934). Later the author described, in rational terms, the Galois groups of maximal ℓ-extensions of class two over $\mathbb{Q}$, with restricted ramification, and applied this description to derive explicit information on class groups of Abelian ℓ-extensions, which went far beyond genus theory. He determined in particular all such extensions with class number prime to ℓ (Fröhlich 1954, 1955). These papers were difficult to read, partly at least because at that time their author only knew the ideal theoretic formulation of class field theory. The comparative ease of the proofs we have given here, bears witness to the superiority of idelic methods in some areas of algebraic number theory.

Important developments in a different direction, also involving central extensions, were initiated by I. R. Safarevic and continued by him and his collaborators. Over the years these led to three major successes: The existence proof for Galois extensions of $\mathbb{Q}$ with given finite soluble Galois group (Safarevic 1954), where A. Scholz had already made substantial progress, the solution of the class field tower problem (Safarevic 1962, Golod-Safarevic 1964) and the explicit determination of the Galois group of the maximal ℓ-extension of a local field (Demuskin [96], see also Serre 1962/63).

Moreover a general theory for Galois groups of ℓ-extensions with restricted ramification was developed - a good source up to 1970 is (Koch 1970).

For the relevant Galois cohomology see Serre's Springer notes (Serre 1969) and Haberland's thesis (1978).

The down to earth explicitness and quantitative precision of the theory over $\mathbb{Q}$, as described in these notes, even now goes far beyond what the general theory can achieve. Its specific source lies in our Theorems 3.12 and 3.13. To my knowledge the latter has been stated here in such a general context for the first time, although in principle one must assume both to have been known. A special case, when the intermediary extension is an Abelian field, is implicit in the author's work on fields of class two and was stated explicitly in connection with a central embedding problem over $\mathbb{Q}$ (Fröhlich 1959). In the general context of projective representations of Galois groups over $\mathbb{Q}$, Theorem 3.12 has been proved by Tate (Serre 1977).

One of the problems which has recently led to a renewal of interest in this general area is that of the validity, or otherwise, of the norm principle, or more precisely the determination of the number knot. A recent fundamental paper, which also gives some background and has a good list of references is (Jehne 1979).

An account of genus theory has been included as necessary background material. The genus field had first been defined by A. Scholz in the work quoted above. The author (Fröhlich 1959) developed the theory for arbitrary extensions of $\mathbb{Q}$, again to obtain class group results (not in fact being aware that the definition goes back to Scholz). This work was generalized to arbitrary number fields as base fields by Y. Furuta (1967). These ideas were subsequently used by various authors to obtain new class group results (see e.g., Halter-Koch 1978, G. Cornell's Thesis 1978, and the survey of Ishida 1976).

Literature

G. Cornell, Genus fields and class groups of number fields (Brown University Thesis, 1978).

S. P. Demuskin, On the maximal p-extension of a local field, Izv. Akad. Nauk SSSR, 25(1961), 329-346.

A. Fröhlich, On fields of class two, Proc. London Math. Soc. 4(1954), 235-256.

________, On the absolute class group of Abelian fields, Journ. London Math. Soc. 29, (1954), 211-217 and 30(1955), 72-80.

________, The Generalization of a Theorem of L. Rédei, Quart. J. Math. Oxford, 5(1954), 130-140.

________, The Rational Characterization of Certain Sets of Relatively Abelian Extensions, Phil. Trans. Royal Soc., 251(1959), 385-425.

________, The genus field and genus group in finite number fields, Mathematika 6(1959), 40-46 and 142-146.

E. S. Golod and I. R. Safarevic, On the class field tower, Izv. Akad. Nauk SSSR 28(1964), 261-272, English translation Am. Math. Soc. Transl. (2) 48, 91-102.

Y. Furuta, The genus field and genus number in algebraic number fields, Nagoya Math. J. 29(1967), 281-285.

K. Haberland, Galois Cohomology of algebraic number fields, VEB Deutscher Verlag der Wissenschaften, Berlin, 1978.

F. Halter-Koch, Eine allgemeine Geschlechtertheorie und ihre Anwendung auf Teilbarkeitsaussagen für Klassenzahlen algebraischer Zahlkörper, Math. Ann. 233(1978), 55-63.

M. Ishida, The genus fields of algebraic number fields, Springer Lecture Notes, 555(1976).

W. Jehne, On knots in algebraic number theory, Crelle 311/312 (1979), 215-254.

H. Koch, Galoissche Theorie der p-Erweiterungen, Springer, Berlin 1970.

L. Rédei and H. Reichardt, Die durch vier teilbaren Invarianten der Klassengruppe der quadratischen Zahlkörper, Crelle 170(1933), 69-74.

L. Rédei, Arithmetischer Beweis der Satznes über die Anzahl der durch vier teilbarem Invarianten der absoluten Klassengruppe im quadratischen Zahlkörper, Crelle 171(1934), 55-61.

L. Rédei, Über die Grundeinheit und die durch acht teilbaren Invarianten der absoluten Klassengruppe im quadratischen Zahlkörper, Crelle 171(1934), 131-148.

I. R. Safarevic, Algebraic number fields, Proc. Int. Math. Congr: Stockholm 1962, 163-176, English translation Am. Math. Soc. Transl. (2), 31, 25-39.

__________, Construction of fields of algebraic numbers with given solvable Galois group, Izv. Akad. Nauk SSSR, 18(1954), 525-578. English translation Am. Math. Soc. Transl. (2) 4, 185-237, (1956).

A. Scholz, Totale Normenreste, die keine Normen sind, als Erzeuger nicht abelscher Körpererweiterungen, Crelle 175(1936), 100-107 and Crelle 182 (1940), 217-234.

J.-P. Serre, Structure de certains pro-p-groups, Sem. Bowbaki 1962/63, No. 252.

__________, Cohomologie Galoisienne, Springer Lecture Notes 5, 1964.

__________, Modular forms of weight one and Galois representation, Durham Proceedings, Ac. Press 1977.